Eliminating Serious Injury and Death from Road Transport

A Crisis of Complacency

Eliminating Serious Injury and Death from Road Transport

A Crisis of Complacency

Ian Ronald Johnston
Carlyn Muir
Eric William Howard

CRC Press
Taylor & Francis Group
Boca Raton London New York

CRC Press is an imprint of the
Taylor & Francis Group, an **informa** business

CRC Press
Taylor & Francis Group
6000 Broken Sound Parkway NW, Suite 300
Boca Raton, FL 33487-2742

First issued in paperback 2017

© 2014 by Taylor & Francis Group, LLC
CRC Press is an imprint of Taylor & Francis Group, an Informa business

No claim to original U.S. Government works
Version Date: 20131014

ISBN 13: 978-1-138-07713-3 (pbk)
ISBN 13: 978-1-4822-0825-2 (hbk)

Visit the Taylor & Francis Web site at
http://www.taylorandfrancis.com

and the CRC Press Web site at
http://www.crcpress.com

Every century comes with a major public health warning about the harm we inflict on ourselves. In Britain in the nineteenth century it was the diseases we spread by tolerating open sewers. In the twentieth century it was tobacco that we slowly learnt to love then fear. In the twenty-first century it is the way we tolerate how cars are allowed to travel on our roads.

—Danny Dorling, The 21st Westminster Lecture on Transport Safety,
November 2010

Contents

Preface

Both Ian Johnston and Eric Howard have spent lengthy careers in the traffic safety field attempting to translate knowledge gained from traffic safety research findings and hard-won frontline experience into policy and practice, and have long been battling to understand why traffic safety progress lags so far behind what scientific knowledge demonstrates is achievable. The motivation for this book was a desire to make sense of our experiences and frustrations and to set out our conclusions in the hope that we may catalyse a community demand for transformational change, and that we may guide and motivate the future efforts of the myriad others working in the field.

Many, if not most, Western motorised nations regularly celebrate their ongoing improvements in traffic safety: road crash fatality rates per million kilometres driven and per head of population are indeed at historic lows. Sadly, though, the Western focus is invariably on the incremental gains made, and not on the sum of the ongoing losses. We never acknowledge the frightening total of serious injuries and deaths from traffic crashes that we are prepared to tolerate, that we accept implicitly through the targets set in our official safety strategies. Road use is the highest-risk daily activity we engage in, and in this book we argue that collateral damage from daily road use is no longer acceptable, and that future gains must be fundamental rather than marginally incremental.

We explore why, as motorised societies, we are complacent about road trauma and set out our view of what needs to change if we are to evolve a safer road-use culture. The idea for a book addressing why society—and its political leaders—accepts collateral damage from its road transport system grew gradually from our frequent confrontations of the widespread, strongly held view that a (sizeable) quantum of trauma is an inevitable outcome of our highly prized daily road use. The vexed issue of speed behaviour and speed management was the specific catalyst: despite undeniable physics and indisputable research results, speed management remains relatively ineffective.

Ian was finally encouraged to put pen to paper as a result of favourable reactions to his many keynote presentations on these big-picture issues to conferences in Australia, New Zealand, and the United States. He enlisted the help of Eric and Carlyn, and together, over the next 5 years, this book slowly came into being. We expect that many of the ideas will be resisted by vested interest groups, but our hope is that they will inspire a new commitment among influential citizens and leaders and develop as an antidote to the complacency that typifies official approaches to road trauma.

Acknowledgements

The Transport Accident Commission (TAC), the "no fault" transport injury compensation insurer of the government of Victoria, Australia, provided a very generous, untied grant to assist with the preparation of this book. Carlyn Muir is a research fellow at the Monash University Accident Research Centre (MUARC) who has worked in public safety research for a number of years, and her time working on the book was partially funded by this grant.

We are deeply indebted to the TAC not only for their financial contribution to MUARC to finance Dr. Muir's time, but also for introducing us to selected clients of theirs, recipients of injury compensation who were willing to discuss their experiences in the belief that some good may come from what they have endured and continue to endure. We are indebted to these families for agreeing to relive their horrific tragedies.

We also acknowledge Emily Kerr's research assistance, which was funded through the TAC grant. Emily's efforts ensured that we thoroughly explored the latest literature and data, and so based the book on the soundest possible scientific footing.

The TAC is internationally recognised as an innovative leader in the field of traffic safety, a rare example of an insurer who has not only invested in injury prevention, but has made its investments across the full range of countermeasures. This book is an example of the TAC's preparedness to facilitate open, independent discussion, in this case through its provision of an untied grant, despite the potential for a degree of controversy to arise around some of the issues addressed in the book. We stress that the TAC has never sought to influence the book's structure or contents—at the outset, during the course of the project, or at the manuscript stage. Apart from providing an unconditional grant, the TAC's only role has been to facilitate access to compensation recipients wishing to publicly share their stories. In no way, therefore, should the TAC be considered as having endorsed, implicitly or explicitly, anything we have written.

Opinions, conclusions, and any errors are entirely the responsibility of the three authors.

About the Authors

Dr. Ian Ronald Johnston is a psychologist with a PhD in human factors. He has worked in the transport field, specialising in transport safety, for over 40 years. He worked in the Australian government's inaugural road safety unit in the 1970s, headed the Victorian government's road safety unit in the 1980s, was managing director of Australia's premier transport research organisation in the 1990s, and was director of the world-renowned Monash University Accident Research Centre until his retirement (from full-time work) in 2006. Dr. Johnston now runs his own consultancy specialising in helping government and industry develop and implement safety strategies. He is also deputy chair of Australia's National Transport Commission and a member of the Core Advisory Group of the World Bank's Global Road Safety Facility. He was a member of Australia's National Road Safety Council for the 3 years of its existence. Dr. Johnston has published extensively in the field and has received several awards for his work, including being made a member of the Order of Australia in 1997.

Dr. Carlyn Muir is a research fellow at the Monash Injury Research Institute, which incorporates the Monash University Accident Research Centre (MUARC). She is a psychologist whose doctoral research examined driver licensing policy for people following brain injury. She has been involved in the development and review of public health policy not only from a research perspective, but also through direct policy implementation within state health services. Dr. Muir's current research involves the design and implementation of a range of injury prevention and public health projects, with a focus on community health and safety, policy, and evaluation. She has published journal articles, book chapters, and government reports across the community safety space.

Since 2006 **Eric William Howard** has operated a strategic road safety consultancy that provides advisory services to clients including the World Bank, Global Road Safety Partnership, AusAID, PIARC, OECD/ITF, international NGOs, and overseas and Australian national and state governments and corporations. Howard was general manager of road safety for VicRoads, the lead role in the Victorian government, for 7 years. He was also a member of the Ministerial Road Safety Council and the National Road Safety Task Force following a career as a local government CEO and senior engineering manager for 24 years. His experience in public health includes 4 years as a board vice president of Mercy Hospital for Women, a leading teaching hospital in Melbourne. Howard chaired the OECD/ITF working group, which developed the "Towards Zero: Ambitious Targets and Safe System Approach" report, published in 2008; he was principal author for the United Nations Road Safety Collaboration–sponsored *Speed Management Manual* published in March 2008; and he was the independent chair of the Western Australian Parliamentarians' Road Safety Reference Group from 2008 to 2011. He chairs a number of road safety task forces for the New South Wales government and he is co-chair of the 33,900—the Australian Road Safety Collaboration.

Explanatory Note

This book uses Australian English spellings throughout. In order that our arguments might flow as an integrated stream, we have chosen to cite our sources of evidence and opinion in notes contained in the references section at the end of the book, with superscript numbers in the text identifying each. This seemed to us preferable to having each individual point dissected and challenged in isolation. Of course we welcome challenge in keeping with our goal of starting a new style of discussion about where traffic safety policy and practice might head if only, as a society, we can broaden the way we think about the problem.

In order to assist readers throughout the world, we have sought to find the most commonly used terms:

- *Road trauma* is used to describe the aggregate of serious injuries and deaths from traffic crashes. The common Australian and New Zealand term *road toll* has been avoided except where part of a direct quote or the title of a cited reference.
- *Traffic safety* is used as the generic term for policy and practice to reduce road trauma, rather than the term *road safety*.
- *Serious (or disabling) injury* means nonfatal injury requiring a hospital admission duration of more than 24 hours and having medium-term or ongoing impacts of either a physical or psychological nature.
- *km/h* (kilometres per hour) is used for vehicle speed except where a direct quote, the immediate context, or a reference title requires *mph* (miles per hour). A conversion table is provided below for ease of reference. The mph equivalents are given to the nearest whole mile.

km/h	40	50	60	70	80	90	100	110
mph	25	31	37	44	50	56	62	68

- The term *hoon*, though we believe it to be of Australian origin, has been retained because of the colourful way in which it portrays a driver who behaves with reckless disregard for others.
- *Cell phone* is used rather than mobile phone.

1 Eliminating Serious Injury and Death from Road Transport Is Not a Pipe Dream

Every one of us uses the roads every day, most of us several times a day. We walk, cycle, and drive as part of our daily lives; this is how we "get around" in order to do all the things we want to do. Roads, and the vehicles they service, are also the dominant means by which the vast majority of the goods we consume, and almost all of the services we use, move from farm, port, factory, or business to warehouse, tradesman's premise, service centre, or retail outlet to our homes. Road use is such a fundamental part of our lives that we take it for granted. A modern, efficient road and road transport system is unquestionably critical to our standard of living. Not surprisingly, though, given the staggering volume of road use and the ever-present opportunity for something to go wrong, crashes are also commonplace daily events.

Thankfully, the vast majority involves only property damage, but a small proportion results in disabling injury, and a smaller proportion still results in death. Though the latter proportions are small, the absolute number of disabling injuries and deaths is tragically large. Many tens of millions are seriously injured or disabled each year. Someone, somewhere in the world, dies about every 25 seconds as the result of a road crash.[1]

Every day in Australia around four people die, close to 50 are seriously injured, and some 500 family, friends, and work colleagues are directly affected as the impacts of these tragedies spread like ripples across a pond.

Trauma is also increasing relative to other lifestyle issues threatening our well-being in the twenty-first century. As the British medical journal the *Lancet* pointed out during the release of the results of the 2010 Global Burden of Disease study: "Since 1970, men and women worldwide have gained slightly more than ten years of life expectancy overall, but they spend more years living with injury and illness."[2]

Yet the *road toll*—a term in common use in Australia for the aggregate total of disabling injury and death—seems generally accepted in Western societies as the price that we must pay for the road use that is so fundamental to our daily lives. The very word *toll* implies a price that must be exacted, viewed as unfortunate (but apparently acceptable) *collateral damage*.[3]

How strange that we are willing to accept, with mind-numbing equanimity, such a level of human suffering while we simultaneously express outrage at the collateral damage that occurs in the current wars in the Middle East, and we express even

1

greater outrage at the anticipated negative impact of the resultant refugees arriving "illegally" on our shores![4]

Leonard Evans, in his 2004 book *Traffic Safety*, wrote: "In a typical month more Americans die in traffic than were killed by the 11 September terrorist attacks on New York and Washington." The aftermath of 9/11 has been a massive global investment in security systems. At the same time, traffic safety investment has remained pitifully small.[5]

Evans also noted: "Since the coming of the automobile in the early days of the twentieth century, more than three million Americans have been killed in traffic crashes, vastly more than the 650,000 American battle deaths in all wars, from the start of the revolutionary war in 1775 through the 2003 war in Iraq."[5] He concluded that traffic safety is a grossly underestimated public health problem.

It is impossible to escape the conclusion that we have the level of trauma that we, as motorised societies, are comfortable with! We are comfortable in part because, as citizens, we are ignorant of the facts. A video clip from the United States presents the results of a "straw poll" of people in the street that tested their knowledge of the number of road crash deaths.[6] Acknowledging that straw polls are totally unscientific, the results are fascinating. Many of those interviewed grossly underestimated the actual number of deaths; most, though, simply had no idea. Asked what a road trauma target should be, the common view was "less than" or "about half" of what it is now, despite "now" being an unknown! When asked what would be an acceptable level of road trauma *among family and friends*, the answer was zero. Asked again, after this personal context had been introduced, what a national safety target should be, many now opted for zero.

This straw poll provides two grounds for optimism: public concern can be raised both by increasing the level of knowledge of the true human and social impact of road crashes and, more cogently, by making the story personal.

We are also comfortable in part because government and industry leaders keep reassuring us that everything that can be done is being done. That reassurance, sadly, is not anchored in reality, for policy and action lag well behind our knowledge of what could be effectively implemented. Governments, in particular, seem to be shying away from "an inconvenient truth," to borrow the title from Al Gore's film about the science around climate change.[7] They appear to focus not on the enormity of the human and social price borne by the community, but on what they perceive as an unaffordable economic cost of the transformational change to road transport system design and operations that traffic safety science findings demonstrate is needed. No doubt they are also frightened of the prospect of civil liability suits should they accept greater accountability for the level of safety. This will prove a much harder hurdle to jump!

There is myriad empirical evidence confirming the level of both public and official complacency surrounding the public health problem of road trauma. Globally, malaria and road crashes result in about equal numbers of deaths, but the United States, for example, spends about 100 times more of its foreign aid on assisting with global prevention of malaria than it does on crash injury prevention.[8,9] Globally, for every $1 spent on road crash injury prevention, around $26 is spent on HIV/AIDS.[10] Four times as much is spent on cancer research as on road safety research in the

United States. Even the amount spent on dental research exceeds that spent on road safety research.[9]

In this book we try to understand why complacency continues to rule, especially given that we now have sufficient scientific knowledge to eliminate the occurrence of disabling injury and death from road crashes without significantly diminishing the vital role that road use plays in our modern lifestyle.

There are, of course, myriad critical issues facing mankind in the twenty-first century—such as climate change, endemic poverty, energy and food supply, and so on. Is there really such a compelling case for traffic safety to compete for political saliency, at least in Western motorised nations? We believe the answer lies in the finding that road use is the highest-risk daily activity we engage in by a factor estimated as high as seven.[11]

This book is not anticar or antiroads, and we do not see the "solution" as less road travel per person. On the contrary, we embrace the enormous benefits modern road transport has conferred on society. We simply believe that these benefits can be retained without paying the current price in human suffering. Of course, there will be a substantial financial cost to be met to eliminate serious injury and death, but the costs of implementing public policies are something governments have continuously to weigh in deciding how best to expend our scarce shared resources. But, for so long as the community continues to view road trauma as unavoidable collateral damage, governments and industry will be *permitted to continue to limit their preventive efforts*. If we are to make our roads genuinely safe, the community must demand more; a vigorous constituency for safety must build and become heard. Without a public constituency, there can be no political saliency.

This book has been written, in the spirit of the words of Levitt and Dubner, "to start a conversation, not to have the last word."[12]

This is not a technical book about traffic safety. There are many excellent books that provide an accounting of the science, of what we know about causal factors and effective interventions, either specifically with regard to human factors or more broadly across the field of traffic safety.[5,13,14] However, they do not, other than in passing, address the relationships between the science, traffic safety policy, road use culture, and why things do or do not find their way into policy and practice.

This book focuses on why public policy lags so far behind the evidence base, on how this gap might be overcome, and on why it must be overcome if we are to make long-term gains of the kind that the evidence suggests are "there for the taking."

We have written not (so much) for the traffic safety professional but primarily for the intelligent layman—for that surprisingly large proportion of the population who have been "touched by the road toll" (Figure 1.1), for road crash victim support groups, and for media editors and transport and public health reporters. These are the people who will help to catalyse the creation of a vibrant and vocal constituency for traffic safety.

We have also written for politicians at local, state, and national levels. And for senior staff—the information gatekeepers—in agencies whose activities have a bearing on traffic safety but for whom it is not their primary accountability—for example, in road and transport agencies, in injury compensation agencies, and in police, public health, and education agencies. Finally, we have written for road, traffic, and vehicle

FIGURE 1.1 "Touched by the Road Toll" bumper sticker. These bumper stickers were introduced by Victoria's injury compensation insurer—the Transport Accident Commission (TAC)—in an attempt to raise awareness of the extent of road trauma. (Courtesy of TAC Australia.)

engineers whose work impacts fundamentally upon safety but who (collectively) do not have a deep understanding of the "big picture."

These are the many groups of people who need to understand why complacency rules, why a constituency for a truly safe road transport system needs to arise, why it should be embraced, and why they should become leaders in its development.

2 Serious Crashes Happen to Real People

The specific crashes that we see, hear, or read about in the media are those judged *newsworthy*—those with the most tragic outcomes and those involving culpable, reprehensible behaviour, especially if they involve a group of teenagers. Newsworthy typically means crashes that conform with our stereotype of fault on the part of one or more of those involved, and we later explore in some detail how this stereotype evolved and the powerful influence it has had on traffic safety policy and practice, and how it helps explain why complacency rules.

This stereotype is reinforced by the recent popularity of crash investigation shows on television. These usually take the form of a "countdown," where we are asked to digest all the events in the minutes that tick by toward the moment of a crash. This leaves us with the impression that somehow the crash was inevitable; those "seconds until impact" seem predetermined and unavoidable.

All too often we focus on specific illegal behaviours like drunk driving or speeding. While we spend so much time blaming the victim, we forget that most crashes, particularly nonfatal crashes, are not caused by deliberate wrongdoing. Rather, they are far more commonly the result of a simple mistake, a lapse in attention, or an error in judgement made by an imperfect human being. We later present compelling evidence that puts to rest the myth of blame as the principal culprit (see Section 6.2), evidence that, sadly, has attracted virtually no media attention.

Traffic safety progress at a macrolevel—beyond individual crashes—is typically reported in the form of brief summary statistics, limited to fatalities; it is usually the running total of road deaths to date compared with the running total to the same point the year before. Even more misleadingly, focus is placed on holiday weekends and periods such as Easter and Christmas, despite research showing that these periods do not have unusually high incidences of serious crashes[15] (see Box 2.1). If there appears to be a trend, particularly one for the worse, this will also attract attention.

In reality, road trauma is something that cannot be predicted at the individual level; we cannot know just who will be involved. For each of us personally, road trauma is unexpected, unpredictable, and life changing. As Mary Schmich famously pointed out in her *Chicago Tribune* column: "The real troubles in your life are apt to be things that never crossed your worried mind; the kind that blindside you at 4 pm on some idle Tuesday."[16]

In short, what the community typically has are simple counts of fatalities and simplistic indications of trends, along with a daily media diet of dramatic and unrepresentative individual events. It is not surprising that the net result is a misunderstanding of the true state of affairs. We are inoculated with apathy through the steady drip, drip of everyday frequency.

BOX 2.1   MEDIA FOCUS ON HOLIDAY PERIODS

Newcastle Herald
    "Hunter Tragedy Lifts Holiday Road Toll"
    Ben Doherty, January 6, 2000
Illawara Mercury
    "National Holiday Road Toll Hits 57"
    January 5, 2000
Sydney Morning Herald
    "Holiday Death Toll Hits 18 as Travellers Enjoy an Easy Run Home"
    Ellen Connolly, January 4, 2000
Herald Sun
    "Holiday Road Toll 'Unacceptable'"
    Shannon Deery, January 4, 2012
Brisbane Times
    "Police Angered by Holiday Road Toll"
    December 27, 2011

Let us start, then, by putting a human face to serious road crashes. What follows are the stories of three families torn apart by road trauma, lives that have been irrevocably changed by the events of what began as an ordinary day in their lives. These are real people and real families, something that we all too often forget when we spend our time blaming the victims. We owe an enormous debt of gratitude to these families for sharing their stories and reliving the horrors they have experienced, and continue to experience.*

2.1   NOEL AND JAN'S STORY

Noel and Jan are typical Australian retirees, who had been eagerly planning their leisurely travels around the Australian countryside with their caravan. Noel had been a dedicated schoolteacher, and Jan had worked in a supermarket. In 1999 they both retired and started enjoying a leisurely pace after working hard in their careers. Both led active and social lifestyles; Noel played golf twice a week and took pride in tending his enormous vegetable patch or pottering in the garden. Jan also enjoyed gardening and had lots of social activities, like dancing. Their initial trip during their retirement was to Europe, and upon returning, they took their first long caravan trip touring Australia for 6 months. After this first trip, they enthusiastically planned a number of other caravan holidays to make sure they captured all the highlights of Australia.

In 2004, Noel and Jan decided to take another trip in their caravan from Victoria to Queensland (well over 2,000 kilometres one way) to visit their son, Simon. At the time, Jan was aged 59 and Noel 61. They wanted to help Simon move house, and to build a few things in the garden for the children. This trip had the added bonus that

---

* We thank the Transport Accident Commission (TAC) (Victoria's injury compensation insurer) for identifying families who were prepared to share their stories.

their third grandchild had just been born, and they wanted to spend some time with the 6-week-old baby. Both Jan and Noel enjoyed the trip thoroughly. They got to spend lots of relaxing time with the family, and even tagged a few extra days on the end, as they were in no rush to get back to Melbourne, their only deadline being to attend a wedding in late February.

February 19, 2004, started off a Thursday just like any other. Noel and Jan packed up the caravan and left their son's place at 6 a.m., ahead of the long but scenic drive back to Melbourne. The sun was shining, a beautiful Queensland day. They stopped for a cup of tea in a nice park to enjoy the views, and had a chat about where they were going to stop for morning tea. The decision was made to head farther down the highway to one of their favourite bakeries in the picturesque town of Goondiwindi. Noel and Jan never made it there.

The road from Toowoomba to Goondiwindi is an undivided highway, with a single lane in either direction, a typical low-cost, rural road design in the vastness of Australia. Around 10:45 a.m., Jan, who was taking her turn driving, noticed a white, B-double truck approaching in her rear vision mirror. Jan was travelling at around 95 km/h, well within the sign-posted speed limit, and remembers thinking that the truck was approaching very fast. And that is when things went wrong very quickly. The truck overtook Jan, by later official estimates, travelling in the region of 120–130 km/h. However, instead of clearing the car, the draft created by the truck pulled the caravan sideways, dragging it alongside their car. The impact to the caravan caused the car to flip three times, and Jan recalls a white post flying toward them and screaming "hang on" to Noel (Figures 2.1 and 2.2). The truck did not stop, and the driver has never been found.

The next thing Jan remembers was opening her eyes to find the car pinned on its side. Both she and Noel were hanging by their seat belts, with Jan's arm pinned

FIGURE 2.1   What remained of the caravan after the crash. (Courtesy of Noel and Jan.)

FIGURE 2.2   The car after it had been removed from the scene. (Courtesy of Noel and Jan.)

outside the car. She reassured Noel: "It's all right, there's people coming." By the time help arrived, it was becoming evident that something was very wrong with Noel—he kept saying he couldn't feel anything. It took some time to get them out of the car, as the car was so badly damaged they had to be cut out through the windscreen. They knew that time was critical for Noel, so the first ambulance rushed him to a local hospital, with Jan closely following.

Noel has no memory of the events that followed. When Jan arrived at the hospital, she was immediately taken for the first of many surgeries on her arm. When she awoke, Noel was already gone—he had been airlifted to a major tertiary hospital in the nearest capital city. The Royal Flying Doctor Service was called for Jan, and she too was transported to Sydney, where Simon and his parents-in-law were anxiously waiting for her. She arrived at 11 p.m., and discovered that Noel had got there around an hour earlier.

Noel had been taken to the intensive care unit (ICU). A tracheotomy was performed (a cut in the windpipe and tubes fitted) to help him breathe. He was unable to talk or move. His memory is very hazy, and he just recalls moments here and there. He remained in the ICU for 9 days, while medical staff worked tirelessly to stabilise his significant spinal injury. Jan also remained in hospital to recover from the extensive surgery on her arm.

When Noel was stabilised, it was decided that the couple should be transferred to their home state of Victoria for continuing treatment. However, before they were able to transfer Noel by air, it was important that he was immobilised to ensure no further damage to his spine. A halo system was fitted, by drilling four titanium screws through the skin on Noel's scalp and into the skull bone. The halo system uses very firm screws, and a chest plate and back plate hold things in place. When the halo was in place, Noel recalls that the only thing he could move were his eyes. The halo remained in place for 16 weeks.

The flight to Melbourne was organised for February 28, 9 days after the crash. Jan remembers both of them being loaded into the plane on their trolleys. It was a tight fit with the two gurneys, a paramedic, and a medic on board. Noel, in his typical humour, dubbed the paramedic Kojak because of his closely shaved head. The flight took many hours, stopping to refuel at a regional hub, where it was 44°C at 7 p.m. Eventually, the couple arrived in Melbourne where Noel was admitted to an acute care hospital, before being transferred into long-term rehabilitation.

Things started to come to Noel at different stages. His drug regime sent him off to a place of strange dreams, and he remembers thinking that a taxi driver was chasing him over a fare, although he had not been in a taxi for years. He recalls feeling very odd, because he felt like parts of his body were moving, like his hand coming up to scratch his face, but soon realised there was no actual movement happening. Noel believes that he came to the realisation shortly before Jan that he would not recover from his spinal injury, that he would never walk or move his arms again. The doctors soon confirmed his fears, and Noel was advised that he has complete quadriplegia. He has no function in his arms or legs, although he does not require a ventilator to breathe and can swallow food.

Noel was in the hospital for many, many months. During this time, Jan was unable to move back home because she couldn't look after herself properly while recovering from her severe arm injuries. She moved in with her sister, who lived over an hour from the hospital where Noel was staying. This was a difficult time for them both. Jan could not drive and was reliant on friends and family to take her to visit Noel. However, she would phone him every day to chat and keep him updated on the local cricket and football scores. Jan remained with her sister for 6 months before returning home.

During this period, their home was being significantly modified in anticipation of Noel's homecoming: bathrooms were widened, lifts to assist Noel in and out of bed were fitted, flower beds were dug up to make way for ramps to be installed. Critically, Noel would not be allowed to return home until air conditioning was installed—he must keep his temperature stable between 20 and 28°C. If the temperature gets too low, his muscles spasm, and if it gets too hot, he can experience dysreflexia, which causes the heart rate to drop and blood pressure to rise, putting him at risk for a stroke.

After a long and gruelling 10 months in rehabilitation, Noel was discharged from the hospital on December 23, 2004, the couple's 37th wedding anniversary. When asked in what way their life had changed the most, Jan said: "It's routine, everything is a routine." Noel's days are very structured, with three sets of carers every day (morning, afternoon, and evening), except for a Saturday when he has two. On a typical day for Noel, his carers arrive at 7 a.m. He gets ready for the day, assisted to the bathroom and to get dressed, and is sitting up in his remote-controlled chair somewhere between 8:30 and 9 a.m. Either Jan or his carers will bring him a cup of tea, which he drinks through a straw, and they often join him for a chat. After his cup of tea, Noel likes to read the paper, so his carers set him up at his specially modified, voice-controlled computer where he reads the paper online. After reading the paper, he might get out to the garden, depending on the weather. Noel must always have someone with him in case the chair breaks down. Jan recalls that they used to go up and down the street, but they haven't done this for a couple of years because the weather in Victoria has been too unpredictable and Noel's temperature must be kept stable.

After this, Noel likes to watch television, different shows depending on the days. Jan, or his carers, will prepare lunch and feed him, although Noel notes (with a twinkle in his eye) that they really need to learn not to feed him like a baby by scooping up the dribbles, because then he ends up with even more food on his chin! After lunch, it is time for exercises. A tilt table has been set up in his garage, and he is strapped to the table in a standing position for 30 minutes to get the circulation moving. The carers then help him with some arm exercises. Two days a week, instead of doing the exercises at home, Noel's carers drive him to a local hydrotherapy pool to do some exercises. It takes two people to assist Noel with the exercises, as it is quite challenging getting him ready for the pool and to dress him afterwards. Jan used to go with them, but is finding it physically very hard these days.

Jan usually prepares dinner, which she feeds to Noel. She comments that she finds it difficult to judge the portions because cooking for Noel is really like cooking for half a person as he doesn't eat much. After dinner, the evening shift carers arrive to prepare Noel for bed and do some leg exercises, and then it is time to retire for the night.

Since the crash, Jan does many things that she never used to around the house. She has learned to fix things, to chop trees, and to give things a "good try" without any help. Jan also has the responsibility for all the housework, as well as caring for Noel. She prepares food and feeds him, brings drinks during the day, organises his medications, and tends to his bags for bladder and bowel control. Jan does not begrudge any of these things for a second, but it is certainly not what she expected from her retirement.

Jan and Noel are proud grandparents, and all of the grandkids love to visit their Poppy. It was difficult for the older children to comprehend that he will never walk again, and this took some getting used to. Since the crash, Jan and Noel's daughter had her first child, Jan and Noel's fourth grandchild. She loves to spend time with Noel, but he will never have the opportunity to pick her up and cuddle her, or take her to the park.

When asked about what they miss most about their life before the crash, Noel talks about not being able to do anything for himself. The only things he can do are breathe, talk, and swallow food. Everything else has to be done for him. For Jan, the thing she misses most is going out. They can't go outside because of the weather variation, and they can't go visiting friends because the chair doesn't fit inside their houses, or can't go up their stairs. Now, they don't even get asked out very often because it is all too hard. They are visited by some friends, but others have dropped off, and they have observed that some people feel uncomfortable around Noel, forgetting that he is the same person as he was before, just in a chair now. There are no more holidays for Jan and Noel, and when their friends and relatives go off travelling in their caravans, it is a painful reminder that they will never enjoy the type of retirement that they had planned. With the help of their care agency, Jan has been provided respite periods where she can take longer holidays, but there is no way for Noel to join her on these trips.

For social activities these days, Jan goes to Scottish country dancing once a week with her sister. Her sister and her husband come over for dinner, and while the ladies are at class, her brother-in-law stays with Noel (or more often naps on the couch!). The local football club has a place reserved for Noel, indoors in front of the glass, so he can watch the games. Occasionally, when the weather is cooperative, they also watch the local cricket.

Jan goes away a couple of times a year for a short break, at which time her sister or daughter will come and stay with Noel. The couple recalled one terrible experience while Jan was on a short break. One of the midday carers called in sick. Unfortunately, the message didn't get passed on, and Noel was left by himself between morning and evening. Of course, with no movement in his arms, Noel had no way of calling anyone for help. This is the only time they have had such an experience; however, quite understandably it was very unsettling for them both, and Jan seems a little nervous about leaving Noel alone.

Noel found acceptance in his situation pretty quickly. With his typical charm and quiet stoicism, he comments that he doesn't have much energy, so there is no point in wasting it on negativity. Instead, he finds the time to smile, laugh, and not dwell on the events of that day. But Noel's extraordinary attitude does not detract from the absolute enormity of how 1 second can completely rewrite the course of your life, and that of your family.

All too often, it is on some random Thursday morning, while you are just cruising along, minding your own business, that tragedy strikes.

2.2 SAM'S STORY

Sam is a gregarious and insightful 21-year-old who loves to have a chat and spend time with his friends. Two years ago, he was a typical teenager. Sam and his mother recall that he was more into the social than the academic side of school, but he certainly put plenty of effort into both. He was planning on taking a year out of his studies to visit some exchange students in Germany with whom he had made friends during their Australian stay, and was in the throes of planning his exciting trip.

March 17, 2009, was a Wednesday. It was St. Patrick's Day, and although not a traditional Australian celebration, it is one that has certainly been embraced by the locals. Sam and a few of his wide circle of friends decided that it was most definitely an appropriate reason for a celebration, and so they headed down to a local Irish pub. Knowing he would be drinking, Sam chose not to drive to the hotel. After an enjoyable day drinking with his friends, Sam made a decision that changed the entire course of his life. He accepted a lift home with a young female friend who was (later found to be) more than three times over the legal limit for blood alcohol. Sam was the front seat passenger, and this was the first time he had been in a car with a drunk driver. It is likely that his own alcohol level clouded his judgement.

The drive home was through steep country roads, and shortly after 3 a.m., the driver lost control of the car at high speed. The car flipped down an embankment, colliding with a tree on Sam's side. There was no other traffic around, but the noise of the impact, and the screaming of the driver and the backseat passenger, woke up people in neighbouring houses who came running out to help (Figure 2.3). The two

FIGURE 2.3   Picture of the car after the crash. (Courtesy of CFA Australia.)

girls in the car were taken by ambulance to the hospital, and miraculously they had escaped significant injury. Quite a feat considering that the driver was later charged with eight driving-related offences, including driving at almost three times the legal limit for a full licence holder (0.149 BAC)—noting that, as a provisional licence holder, her legal limit for blood alcohol was zero—speeding, reckless driving, and negligent driving causing serious injury.

Sam wasn't so lucky. It became apparent very quickly that Sam was critically ill. Fortuitously, a nurse happened to be one of the people that lived in one of the neighbouring houses, and she was having trouble finding a pulse. The emergency services arrived on scene and immediately arranged for Sam to be flown to a major trauma centre. His condition was very serious. At 5 a.m., Sam's mother received a phone call from a surgeon advising her that Sam had been involved in a crash. He had sustained a number of injuries, including fractures to his pelvis, eye sockets, and nose, and several contusions. However, the most catastrophic injury was a severe brain injury, and they needed her permission to operate on Sam to remove a large blood clot, and relieve the pressure in his brain.

When Sam's parents arrived at the hospital, he was in the intensive care unit and had been placed in a medically induced coma where he remained for 5 days. In that initial period, there was a lot of optimism from the medical staff about his potential for a good recovery despite the gravity of his injuries. However, on day 6, Sam's mother recalls how the prognosis started to become more guarded. Despite withdrawal of the drugs that were used to keep him in a medically induced coma, Sam was not showing any signs of consciousness. On day 10, a meeting between Sam's parents and the medical team painted a gloomy picture. The worst possible outcome for someone in Sam's position typically occurs when the nerve connections in the brain are absent on both sides. They said that Sam's connections were completely absent on one side, and severely damaged on the other. It was thought that when Sam came out of his coma—and there was no telling when that would be—he would be in a persistent vegetative state and would require 24-hour nursing care for the rest of his life.

Despite the doctors' guarded prognosis for Sam, he is a fighter. After spending 3 weeks in intensive care, he was moved to the neurosurgical ward, where he remained unresponsive for a further week. The surgery that he needed to allow room for swelling and reduce the pressure on his brain was a craniotomy, which involved removing a large part of his skull that was covered by a sign that read "No Bone." This piece of skull was not replaced for 4 months. A major breakthrough came on day 25, when Sam opened his eyes for the first time. Sam's recovery from the coma was nothing like the movies—there was no big dramatic and tearful scene where he woke up surrounded by loved ones and had an epiphany about his life. In the initial days, Sam only had extension reactions to pain on one side. Very occasionally, he would open his eyes. Instead, Sam's parents realised that recovery was a very slow process and were very quickly taught to lower their expectations. They learned to be happy with very small responses, and to see each one as a little victory. His mother likened it to an iceberg thawing, tiny piece by tiny piece.

One of Sam's major problems was that he had significant posttraumatic amnesia (PTA). PTA is a state of confusion and memory loss that can occur following a traumatic brain injury, and is often considered to be an indication of how severe the brain injury is. It can be anterograde, where the person has problems creating new memories after the injury has happened, or retrograde, where there is difficulty recalling events from just prior to the injury. In Sam's case, he had both. To this day, he has no memory for the period of around 2 weeks before the crash, and limited memory for around 4 months after the crash.

On day 35, Sam was transferred to a rehabilitation hospital more than an hour's drive from his parents' house. He had a tracheotomy performed when he first arrived at hospital, and this made his rehabilitation challenging, to say the least—the staff were limited in what they could do. Sam's mother remembers his responses being really good when he was put on a tilt table. This involved strapping Sam to a table and lifting him into a standing position. During this particular therapy he seemed to be quite responsive and alert, but it was a huge effort for everyone involved because of the number of staff needed to manage him on the table. One nurse was needed just to manage the tracheotomy, and another to hold him in position, while a third nurse monitored his blood pressure and a physiotherapist operated the controls. He also had to wear a rugby helmet at all times to cover the area of the skull where there was no bone.

Frustratingly for Sam, he was unable to speak for months. He could respond to voices by turning his head toward the person speaking, but there was no focus in his eyes. Sam's mother recalls that he had a glassy stare that seemed to look right past you. His parents thought it would be useful for him to try drawing or writing, so they brought him in a sketchbook. This turned out to be a useful outlet for Sam, and he was able to communicate with the outside world. The first thing Sam wrote in his sketchbook was very telling: "Car is scary, God is scary" (Figure 2.4).

Throughout his recovery, Sam had vivid and frightening dreams. He had to be restrained in the hospital because one side of his body would fight against the demons in his dreams, flailing his arms around, pulling out various tubes and wires, and causing problems for his tracheotomy. When he was transferred to the rehabilitation hospital, they decided to try something different rather than restraining him. Instead, he was confined day and night in a CraigBed, which is an enclosed cot with padded walls 2 metres high. Sam recalls one particularly frightening dream where he thought he was jumping onto a boat, and just as he jumped, the boat moved. What Sam had actually done was climb the walls of his bed and jump off the side. And this was at a time when he weighed only 45 kilograms after having significant trouble absorbing nutrients, and he still had a piece of his skull missing. The medical team to this day do not know how he managed to scale the walls in his condition.

Sam's rehabilitation program was lengthy. It consisted initially of physical therapy; he had to be taught how to sit up again, and when the breaks in his pelvis had healed, to stand up and walk. There was also considerable psychological therapy to help with his PTA, as well as physiotherapy, occupational therapy, speech therapy, and music therapy. The music therapy was something that Sam enjoyed because although he couldn't remember a lot of things, he could remember music, and this gave him hope. Sam was committed and cooperative with his rehabilitation, particularly with his physical therapy. His dogged determination to succeed could be seen in every milestone that he reached. His mother notes that he also had his challenges. Understandably, Sam could get frustrated when he couldn't do things he wanted to,

FIGURE 2.4 Sam's first communication after the crash. The text reads: "Car is scary, God is scary, Crashed." (Courtesy of Sam.)

and could behave erratically. However, what is very clear is that despite the odds, Sam was very determined to succeed in his own recovery.

Sam remained in the rehabilitation hospital for more than 4 months. Before he could be discharged home for outpatient therapy, he needed to have some trial nights at home, and for a couple of months prior to his discharge, he would spend one night on the weekend at home getting used to things. Finally, in early September, Sam was allowed to go home. His parents came to the hospital to collect him, and after a final meeting with medical staff, he was officially discharged. Sam remembers this day well; Michael Jackson had recently died, so Sam wore a pair of snakeskin shoes to commemorate his passing. The three of them returned home to a party of excited family and friends, and Sam lasted until 7:30 p.m., when he took himself off to bed.

In the 2 years that have passed since Sam came home from the hospital, his life has changed in so many ways. He has gradually found ways of doing most of the things he did beforehand, albeit having to adapt the way in which he does them. A glaring omission is tennis, one of his favourite pastimes before the crash, but his responses are too slow to play now, and he doesn't have many people to play with. His social life has been significantly impacted by his brain injury, and he has only a few friends who have stuck by his side. In some ways this is because Sam spent such a long time unable to participate in his friends' lives, and in other ways because people find it uncomfortable that Sam is no longer the person he was before the crash. He finds it very difficult to pay attention, and can drift off in conversation or while trying to concentrate on things. His memory can also be a problem, very common following a brain injury.

In 2012, Sam enrolled in a part-time multimedia course, and is enjoying learning new skills. Multimedia has been a great creative outlet for him during his recovery, and is something he can see himself doing in the future, although he has also developed a keen interest in behavioural science and helping people in his situation. Getting back into education has been really helpful for Sam, as it is giving him an opportunity to make new friends and start afresh. This time he is much more focused on his schoolwork, rather than the social side. Sam also realised his dream to go to Germany in 2010, visiting all the places he was dreaming of before the crash. However, this was a trip he took with his parents, which, although thoroughly enjoyable, was not quite how he imagined it would be as he planned it before his crash!

Sam is very insightful about the crash and how it has changed his life. He talks about evolving as a person, and the ripple effect that one decision has had on his life. One thing he is particularly passionate about is making people aware of the devastating effects of road trauma, and he has been involved in several media campaigns to bring it front and centre in people's minds. However, he is also realistic. He comments that when you see someone on TV who has been in a crash, it might have quite an impact on you while you are watching, but it is only in your mind for a few seconds, until the next "commercial break" comes on. Then you forget all about that person, and all the others whose lives have been irrevocably changed by a split second in the aftermath of a crash. It is not just the slow and painful recovery that Sam himself experienced that is difficult to conceive, but the extraordinary toll it took on his family's life too. Sadly, there are no commercial breaks in Sam's life, and no opportunities to forget.

Sam's mother muses that it has been more than 1,000 days since the crash, and each of these days has been more difficult for Sam than any day that he had experienced in his life before. Similarly, each of these days has been the most difficult the family has ever endured.

None of their lives will be the same again, and it all stemmed from a single decision. Sam knew that he should not drive that day and made plans accordingly, but he was unable accurately to judge whether that was also the case for the person who offered him a lift.

2.3   ABBEY'S STORY

Abbey is recovering, slowly but progressively, from a severe brain injury. In February 2011 she was a vibrant and attractive 19-year-old, living life to the fullest in a small country town in Victoria. She ran 5 kilometres twice a day, was an athlete, played cricket and football, was an apprentice builder, and achieved fame as the first female in 75 years to become a volunteer firefighter with the town's brigade.

All that changed just after 4 p.m. on a weekday in February 2011. The day started as an unusually busy one for the family. Abbey's elder brother was moving away from home to be closer to the regional university that he attended, while her mother had that day resumed her teaching position on the first day of the new primary school year. Abbey helped with the packing for her elder brother and later volunteered to pick up her younger teenage brother and his friends from school. Having picked up the group and taken each in turn to their homes (including her younger brother), she arrived at the home of her last passenger, a teenage boy about 3 years her junior. He asked if he could come for "a spin," as Abbey needed to drive to a nearby town to refuel.

These background circumstances were provided by Abbey's parents, as Abbey (like Sam and so many with brain injury) has no recall of the events preceding or for some time after her crash.

The crash occurred on a section of local road near her hometown, a section with which she was very familiar. It was a typical two-lane, two-way rural road and was part of a route Abbey had driven frequently with her father while she was a learner driver acquiring the 120 hours of supervised experience required before a Victorian learner can apply for a provisional driving licence. Her parents had been conscientious in ensuring that she logged her quota of hours under a variety of driving conditions.

Abbey had graduated to the second stage of her provisional licence a short time before her crash. (In Victoria, the first 12 months is spent displaying a red P-plate—with several restrictions—and the rest of the provisional period is spent displaying a green P-plate with fewer restrictions.) Her father regarded her as a competent driver, and there is no suggestion that she regularly exceeded the speed limit.

Abbey and her young passenger drove this familiar route after completing the refuelling. The route included what is locally known as the "big dipper"—named after amusement park rides—where two short, steep hills occur in succession. Abbey lost control as she crested the second of these hills, and her car veered off the road to the right, impacted a culvert, and cleared a fence of over a metre in height, rolling several times (Figure 2.5).

The local impact on the community was substantial. The local policeman, known to the family, held Abbey's hand while her colleagues from her local volunteer fire brigade helped release her from her trapped position in the car. Her mother arrived on the scene while the recovery process was still underway, and her father soon after. There was none of the victim anonymity that helps shield emergency services folk in large urban areas, and her parents experienced, firsthand, the full horror of the crash.

There were serious concerns that Abbey may not survive. Her airbag had not activated (Figure 2.6), a failure that almost certainly contributed to the severity of her injuries. The vehicle was a common, large station sedan, about 10 years old. Abbey's pelvis was broken in two places, and she had a major head injury. Her passenger's airbag did activate, and he received relatively minor injuries.

The police report states that the road was dry, the weather clear, and the skid marks extended for 70 metres. The engine and forward suspension had separated from the vehicle. The police estimate of precrash speed was 170 km/h, an estimate Abbey's parents dispute. Abbey's mother subsequently drove the same route at the speed limit to measure the time she took between the service station and the crash scene to compare it with Abbey's time computed from a cell phone record of a call from Abbey made at the time of refuelling to the time of the crash reported by witnesses. The times correlated closely. While the bulk of the trip may have been made at or close to the speed limit, there can be little doubt that the precrash travel speed over the second crest was well above the posted limit of 100 km/h. Speeding, it

FIGURE 2.5   The debris strewn across the paddock after the crash. (Courtesy of Abbey.)

FIGURE 2.6   View of the driver's side after the crash. Note that the airbag did not deploy. (Courtesy of Abbey.)

would appear, was momentary, possibly confined to the big dipper section of road. What influence, if any, Abbey's teenage male passenger had on her apparently atypical behaviour is unknown.

Abbey was placed in an induced coma at the scene and transported by road ambulance to Melbourne some 200 kilometres distant. She remained in an induced coma for 19 days while attempts were made to reduce the high intercranial pressure stemming from her diffused axial brain injury. When Abbey was first brought out of her coma she had a severely congested chest and needed a tracheotomy to clear it, and so had to be placed back into the coma for a further period. The stress on her family was enormous—for a long time they had not known whether she would live or die, and then when it appeared that she would live, hope was almost immediately dashed. Abbey's mother's personal diary notes: "The roller coaster has made me cry many times today. I have a sore eye."

The intensive care unit that Abbey was placed in was, coincidentally, the same unit that Abbey's uncle (her father's brother) had been in when he died, and her mother tried vigorously to have Abbey moved lest the coincidence be a portent of more tragedy. The hospital finally convinced her that Abbey's best chances of survival were in this very unit.

Abbey remained in the hospital for almost 8 months, with the final 6 weeks being in a live-in centre preparing to return home, which she visited on weekends. With the hospital so far from home, Abbey's mother was off work for the nearly 4 months that she spent in Melbourne. At first she lived with her twin sister, but the pressure she was under created a rift from which she had to withdraw. (Fortunately, the rift has proved temporary.) Abbey's father spent the first 3 weeks in Melbourne before having to return to work in his hometown. He is full of praise for the compassion shown by his employer.

Abbey's mother kept a personal, daily diary as part of her coping mechanism, but also made daily entries on Abbey's Facebook page. This was crucial, as it served as a reliable source of information for Abbey's friends in the small country town where rumour was rife and other Facebook pages were suggesting, for example, that Abbey's life support systems were about to be shut down!

How has Abbey's life changed? The most immediate change was that she would often ramble in conversation and frequently become emotional and offensive. Such behaviour is now less and less common. Of longer-term impact is the loss of her once wide circle of friends. People of her age love to party, and many experiment with alcohol and other substances. Abbey cannot party and can take only the occasional glass of alcohol. She now has only a small circle of close friends.

A critical change has been a (relative) loss of judgement about what is right and what is wrong and knowing who to trust and who not to trust. As her recovery progressed her parents began to allow her to attend parties, but some boys tried to slip her drinks and take advantage of her relatively poor judgement. In Abbey's terms, the biggest change in her life is the loss of a normal round of socialising and "growing up." For her parents, now that the recovery is proceeding well, the biggest change is their inability to get away for extended holidays, and this will last as long as it takes for Abbey's judgement of the degree of trust she should place in various people to improve to its former level.

The thing Abbey now wants most is to return to work and recommence a normal life—after more than 12 months lost. She has commenced a work trial waitressing (a few hours a week) at a hotel in a nearby town where she had worked casually before her crash. She also hopes to sit a driving test to regain her licence as soon as the neurological reports suggest that her recovery is sufficiently advanced. She is quite positive and very determined, and with a little luck, she will return to a near-normal life, if not one as vibrant as her precrash profile suggested might unfold for her.

One hope that Abbey and her family cherish is that sharing her story may help save other young people from similar tragedy. The hospital in which Abbey spent so much time is one of four in Australia where the Prevent Alcohol and Risk-Related Trauma in Youth (PARTY) program operates. PARTY is a full-day education program exposing senior school students (and, in one variant, young driving offenders) to the consequences of serious crashes. PARTY began in Canada and now operates in over 100 hospitals worldwide. Abbey was a "patient encounter" for three groups of senior school students, and her mother recalls how potent the safety message was for students confronted with a seriously injured person so alike in age and background to many of them.

We trust that the hope of Abbey and her family that some lasting good can come from her crash will be realised.

2.4   THREE STORIES AMONG TENS OF THOUSANDS

For every person who dies in a road crash, more than a dozen are seriously injured, yet Western societies continue to focus primarily on the number of deaths—the tip of the trauma iceberg. We could fill an entire book with stories like those of Noel and Jan,

Sam, and Abbey—ordinary Australians whose lives have been irrevocably changed by events that are frighteningly common. Jan was driving perfectly safely; Sam had been drinking but knew better than to get behind the wheel himself; and Abbey had a momentary lapse in her normally safe driving style. None fits the stereotype that we so often see in the media and hear repeated by far too many of our politicians.

We have to try to understand why such a high level of road trauma is acceptable within our society if we are to progress beyond incremental improvements in the level of safety within our road and road transport system.

Let us hope that these three stories help create a climate of enduring concern that we see as critical to transformational change.

3 The Way We View Safety Is a Big Part of the Problem

3.1 INTERNATIONAL CONCERN IS FOCUSSED ON THE MOTORISING WORLD, NOT ON "US"

The 2009 World Health Organisation's (WHO) *Global Status Report on Road Safety* estimated that more than 1.2 million people die each year on the world's roads, and that between 20 and 50 million suffer nonfatal injuries.[17] It also estimated that more than 90% of the world's fatalities resulting from road crashes occur in low-income and middle-income countries, which have only 48% of the world's registered motor vehicles. Of immense concern is the fact that these countries are now rapidly motorising and the death and disabling injury totals will increase dramatically over the next couple of decades unless radical change occurs.

> Further, (in these countries) since road traffic injuries affect mainly males (73 percent of deaths), and those between 15 and 44 years old, this burden is creating enormous economic hardship due to the loss of family breadwinners.[18]

This hidden effect of road trauma is a major threat to the international community's millennium goal of eliminating global poverty. In its 2012 status report, the WHO concluded that while road traffic injuries were the eighth leading cause of death globally, they were the leading cause of death for those aged 15–29 years.[17]

The forecast explosion in death and disabling injury from road traffic use as motorisation spreads ever more rapidly in the first quarter of the twenty-first century has generated international action, for the first time, beginning with the WHO world report on traffic safety in 2004.[19] The Make Roads Safe Foundation[20] later took action to support an International Ministerial Conference on Road Safety, hosted by the Russian Government in Moscow in 2009, which was followed by a formal United Nations declaration of a "Decade of Action for Road Safety" launched in May 2011 for the period through to 2020 (see Box 3.1).

The UN Decade of Action is a commitment to contain the predicted global growth in road crash deaths by 2020 to half of that currently forecast. The difference between an unchanged global trend in fatalities and serious injuries and a 50% reduction from the forecast over the decade is the avoidance of 5 million deaths and 50 million injuries.

This is unquestionably a laudable goal, but even if achieved, there will be, in 2020, almost the same absolute number of serious injuries and deaths in the world as now!

BOX 3.1   THE MOSCOW DECLARATION[21]

FIRST GLOBAL MINISTERIAL CONFERENCE ON ROAD SAFETY: TIME FOR ACTION

MOSCOW, 19–20 NOVEMBER 2009

MOSCOW DECLARATION

We, the Ministers and heads of delegations as well as representatives of international, regional and sub-regional governmental and nongovernmental organizations and private bodies gathered in Moscow, Russian Federation, from 19–20 November 2009 for the First Global Ministerial Conference on Road Safety,

Acknowledging the leadership of the Government of the Russian Federation in preparing and hosting this First Global Ministerial Conference on Road Safety and the leadership of the Government of the Sultanate of Oman in leading the process for adoption of related United Nations General Assembly resolutions,

Aware that as described in the 2004 World Health Organization/World Bank *World report on road traffic injury prevention* and subsequent publications, road traffic injuries are a major public health problem and leading cause of death and injury around the world and that road crashes kill more than 1.2 million people and injure or disable as many as 50 million a year, placing road traffic crashes as the leading cause of death for children and young people aged 5–29 years,

Concerned that more than 90% of road traffic deaths occur in low-income and middle-income countries and that in these countries the most vulnerable are pedestrians, cyclists, users of motorised two- and three-wheelers and passengers on unsafe public transport,

Conscious that in addition to the enormous suffering caused by road traffic deaths and injuries to victims and their families, the annual cost of road traffic injuries in low-income and middle-income countries runs to over USD 65 billion exceeding the total amount received in development assistance and representing 1–1.5% of gross national product, thus affecting the sustainable development of countries,

Convinced that without appropriate action the problem will only worsen in the future when, according to projections, by the year 2020 road traffic deaths will become one of the leading causes of death particularly for low-income and middle-income countries,

Underlining that the reasons for road traffic deaths and injuries and their consequences are known and can be prevented and that these reasons include inappropriate and excessive speeding; drinking and driving; failure to appropriately use seat-belts, child restraints, helmets and other safety equipment;

the use of vehicles that are old, poorly maintained or lacking safety features; poorly designed or insufficiently maintained road infrastructure, in particular infrastructure which fails to protect pedestrians; poor or unsafe public transportation systems; lack of or insufficient enforcement of traffic legislation; lack of political awareness and lack of adequate trauma care and rehabilitation,

Recognizing that a large proportion of road traffic deaths and injuries occur in the context of professional activities, and that a contribution can be made to road safety by implementing fleet safety measures,

Aware that over the last thirty years many high-income countries have achieved substantial reductions in road traffic deaths and injuries through sustained commitment to well-targeted, evidence-based injury prevention programmes, and that with further effort, fatality free road transport networks are increasingly feasible, and that high-income countries should, therefore, continue to establish and achieve ambitious road casualty reduction targets, and support global exchange of good practices in road injury prevention,

Recognizing the efforts made by some low- and middle-income countries to implement best practices, set ambitious targets and monitor road traffic fatalities,

Acknowledging the work of the United Nations system, in particular the long standing work of the United Nations Regional Commissions and the leadership of the World Health Organization, to advocate for greater political commitment to road safety, increase road safety activities, promote best practices, and coordinate road safety issues within the United Nations system,

Also acknowledging the progress of the United Nations Road Safety Collaboration as a consultative mechanism whose members are committed to road safety and whose activities include providing governments and civil society with guidance on good practice to support action to tackle major road safety risk factors,

Acknowledging the work of other stakeholders, including intergovernmental agencies; regional financial institutions, nongovernmental and civil society organizations, and other private bodies,

Acknowledging the role of the Global Road Safety Facility established by the World Bank as the first funding mechanism to support capacity building and provide technical support for road safety at global, regional and country levels,

Acknowledging the report of the Commission for Global Road Safety *Make roads safe: a new priority for sustainable development* which links road safety with sustainable development and calls for increased resources and a new commitment to road infrastructure safety assessment,

Acknowledging the findings of the report of the International Transport Forum and the Organisation for Economic Co-operation and Development *Towards*

zero: ambitious road safety targets and the safe system approach and its recommendation that all countries regardless of their level of road safety performance move to a safe system approach to achieve ambitious targets,

Acknowledging the findings of the World Health Organization/UNICEF *World report on child injury prevention* which identifies road traffic injuries as the leading cause of all unintentional injuries to children and describes the physical and developmental characteristics which place children at particular risk,

Recognizing that the solution to the global road safety crisis can only be implemented through multi-sectoral collaboration and partnerships among all concerned in both public and private sectors, with the involvement of civil society,

Recognizing that road safety is a 'cross cutting' issue which can contribute significantly to the achievement of the Millennium Development Goals and that capacity building in road traffic injury prevention should be fully integrated into national development strategies for transport, environment and health, and supported by multilateral and bilateral institutions through a better aligned, effective, and harmonized aid effort,

Conscious that global results are the effect of national and local measures and that effective actions to improve global road safety require strong political will, commitment and resources at all levels: national and sub-national, regional and global,

Welcoming the World Health Organization's *Global status report on road safety*—the first country by country assessment at global level—which identifies gaps and sets a baseline to measure future progress,

Also welcoming the results of the projects implemented by the United Nations regional commissions to assist low-income and middle-income countries in setting their own road traffic casualty reduction targets, as well as regional targets,

Determined to build on existing successes and learn from past experiences,

Hereby resolve to

1. Encourage the implementation of the recommendations of the *World report on road traffic injury prevention,*
2. Reinforce governmental leadership and guidance in road safety, including by designating or strengthening lead agencies and related coordination mechanisms at national or sub-national level;
3. Set ambitious yet feasible national road traffic casualty reduction targets that are clearly linked to planned investments and policy initiatives and mobilize the necessary resources to enable effective and sustainable implementation to achieve targets in the framework of a safe systems approach;
4. Make particular efforts to develop and implement policies and infrastructure solutions to protect all road users in particular those who are

most vulnerable such as pedestrians, cyclists, motorcyclists and users of unsafe public transport, as well as children, the elderly and people living with disabilities;

5. Begin to implement safer and more sustainable transportation, including through land-use planning initiatives and by encouraging alternative forms of transportation;

6. Promote harmonization of road safety and vehicle safety regulations and good practices through the implementation of relevant United Nations resolutions and instruments and the series of manuals issued by the United Nations Road Safety Collaboration;

7. Strengthen or maintain enforcement and awareness of existing legislation and where needed improve legislation and vehicle and driver registration systems using appropriate international standards;

8. Encourage organizations to contribute actively to improving work-related road safety through adopting the use of best practices in fleet management;

9. Encourage collaborative action by fostering cooperation between relevant entities of public administrations, organizations of the United Nations system, private and public sectors, and with civil society;

10. Improve national data collection and comparability at the international level, including by adopting the standard definition of a road death as any person killed immediately or dying within 30 days as a result of a road traffic crash and standard definitions of injury; and facilitating international cooperation to develop reliable and harmonized data systems;

11. Strengthen the provision of prehospital and hospital trauma care, rehabilitation services and social reintegration through the implementation of appropriate legislation, development of human capacity and improvement of access to health care so as to ensure the timely and effective delivery to those in need;

Invite the United Nations General Assembly to declare the decade 2011–2020 as the "Decade of Action for Road Safety" with a goal to stabilize and then reduce the forecast level of global road deaths by 2020;

Decide to evaluate progress five years following the First Global Ministerial Conference on Road Safety;

Invite the international donor community to provide additional funding in support of global, regional and country road safety, especially in low- and middle-income countries; and

Invite the UN General Assembly to assent to the contents of this declaration.

Moscow, Russian Federation
20 November 2009

While it is a tough goal just to limit the growth, it leaves absolutely unaddressed the question of why we are prepared to tolerate massive levels of road trauma now and well into the future. It reinforces our statement in Chapter 1 that we, as communities, accept the principle of collateral damage as the price for universal motorisation.

Sadly, the adoption of even this growth containment goal is far from widespread at national government level among low- and middle-income countries. The 2009 WHO *Global Status Report on Road Safety* reported that only one-third of countries have a national road safety strategy that both includes specific numerical targets and has appropriate management systems and funding allocated for its implementation. While the Decade of Action has sparked the adoption of targets in many countries, few targets are based on quantitative analysis, and most seem to have been adopted in "hope" rather than with meaningful commitment.

It is not difficult to document ambition; it is another matter entirely to plan and implement a strategy to realise it! As the decade rolls on, there is a growing concern that what is being done is constrained by governments at the national level by what are seen to be competing road transport objectives and an inability to tackle cultural barriers to the development of a truly safe system. As a commentator noted in response to the 2013 global status report: "Road safety doesn't have political or financial support to move it into [the] mainstream of developmental work."[22]

It is abundantly clear that the Decade of Action is directed at low- and middle-income motorising countries. Motorised nations continue to be complacent about where they are at, and where they are heading. As we will see later, the official forecast is for a further 30% fall in road crash deaths in the West before 2020, so international concern is focussed entirely on what happens in nations where increases are forecast.

It seems we are content with where we—the motorised countries—are heading. But the reality is that 30% fewer deaths by 2020 in motorised nations means that there will still be a huge number of deaths between now and 2020, and well beyond. Our contention is that where motorised nations are at now, and where they are planning to get to in the next decade, is tantamount to official acceptance of an enormous "sacrifice" of lives—a staggering level of collateral damage—during the term of current strategies (see Table 3.1).

For this reason this book focusses on *Western motorised nations* and the changes *we* need to make. At the same time, we acknowledge and wholeheartedly support the Decade of Action; if motorising nations adapt and adopt the successful measures we have developed, the growth in global road trauma can be slowed.

3.2   SHOULD WE MEASURE SAFETY AS ACTUAL NUMBERS OR AS RATES?

How should we measure traffic safety? It all depends upon your view of where traffic safety sits in the bigger scheme of things. If you think disabling injury and death are, in the terms favoured by economists, *negative externalities* of road transport—the unintended but unavoidable downside of an essential system—then you will almost certainly focus on rates, not on the actual numbers.

TABLE 3.1
The Number of Lives to Be Sacrificed

Country	Target in Current National Strategy	Conservative Calculation of Deaths during Strategy (assumes targets achieved immediately)	Implied Sacrifice between Now and Target Date
Australia	Target of (at least) 30% reduction in deaths and serious injuries by 2020	10 years (2011–2020) × 1,292 fatalities (2011 figures) × 30%	12,920 lives
Canada	Achieve 2,500 or less fatalities per year by 2028	18 years (2010–2028) × 2,500 per year	40,000 lives
Denmark	Target of 40% reduction in fatalities by 2012	3 years (2010–2012) × 303 fatalities (2009 figures) × 40%	545 lives
Greece	Reduce fatalities by 50% by 2020	10 years (2011–2020) × 1,300 fatalities × 50%	6,500 lives
Japan	Reduce fatalities to 5,000 or less by 2012	3 years (2010–2012) × 5,000 fatalities	15,000 lives
Malaysia	Reduce fatalities to 10 per 100,000 population by 2020	11 years (2010–2020) × 10 × 264 (26.4 million population)	29,040 lives
The Netherlands	Target by 2020 is to reduce fatalities and serious injuries by a third	11 years (2010–2020) × 644 fatalities (2009 figures) × 30%	4,959 lives
New Zealand	Long-term reduction of severe injuries and fatalities by half	10 years (2011–2020) × 384 fatalities (2009 figures) × 50%	1,920 lives
South Africa	Reduce fatalities by 50% by 2015 (less than 7,000 fatalities per year)—draft target	5 years (2011–2015) × 7,000 fatalities	35,000 lives
Sweden	Interim target for 2020 is 50% less fatalities	11 years (2010–2020) × 358 fatalities (2009 figures) × 50%	1,969 lives
Switzerland	Proposed program to achieve long-term reduction of fatalities by half	2010–2020 (11 years) × 349 fatalities (2009 figures) × 50%	1,920 lives

Source: Bliss, T., and J. Breen, *Implementing the Recommendations of the World Report on Traffic Injury Prevention*, World Bank, Washington, DC, 2009; PIARC, *Comparison of National Road Safety Policies and Plans*, PIARC Technical Committee C.2 Safer Road Operations, 2012.

If everybody needs to use the roads every day (as they do), and if an efficient road transport system is critical to a modern economy (as it is),[25] then the economists' argue that you want the safest system that cost-benefit analysis suggests is reasonable. In short, you will balance the economic benefits of proposals for change against the costs required to achieve that change. You will put an economic price on injury and death and enter this into your calculation.

In our view, the economist's term *negative externality* is the equivalent of the military's term *collateral damage*. Try getting Noel or Sam or Abbey, or any member of their families, to accept they are simply part of the price society must pay! Our view is that human life is sacrosanct and the notion of collateral damage is unacceptable. There is no one we know who accepts the notion of collateral damage as the price we must pay for affordable air or rail transport. Everyone welcomes cheaper and cheaper air and rail fares, but no one accepts that these should occur at the cost of safety. These modes aim for zero disabling injury and death.

Why, then, do we accept collateral damage in road transport? The main reason seems to turn on our beliefs about risk control. We know that, as individuals, we have no personal, direct control over safety in air or rail travel, while we (mistakenly) believe that we have total direct control in our everyday road use. If society's safety objective were to become that of eliminating collateral damage from our road transport system, then our search for effective solutions and our questioning of our fundamental system design and operating assumptions would look quite different.

We certainly have a strange way of viewing safety in road transport. When we say *safe*, we officially mean accepting a level of risk as low as achievable at reasonable cost. Superficially, this seems akin to the ALARP (as low as reasonably practical) principle adopted for industrial safety. But in industrial safety we construct formal risk matrices where we identify all the potential risks and systematically examine both probability of occurrence and severity of consequences given occurrence. No such risk matrix is constructed as a base for road transport system design and operation; traffic safety policy depends upon an institutional assessment of the cost of proposed interventions against funds currently available, mediated by community acceptability. For example, single-vehicle run-off-road crashes are the single most common form of fatal crash on roads in rural Australia—so they are of both high probability and high severity—but our remedial action is patchy at best.[26]

International benchmarking, a popular pastime among national governments, depends upon rates for its comparisons because absolute numbers ignore differences in population size. How, for example, can you compare the 266 road crash deaths in Sweden in 2010 with the 32,885 road crash deaths in the United States in the same year? Rates (per head of population or per unit of road use) are, in this important sense, an inescapable element of assessing progress and performance. The downside of rates, especially when used to set targets, is the assumption that a finite level of trauma is inevitable. It also means that governments set targets to achieve a level "at least as safe as" someone else or "safer than we used to be," both of which are not only rate-dependent terms but also targets that accept that a level of collateral damage is a given.[27]

Historically, we have treated disabling injury and death as negative externalities of our road transport system, but there are signs that change is at least in the wind.

The Swedish parliament, as long ago as 1997, passed a motion to the effect that no one behaving reasonably during their road use should be disabled or killed as a result of that use.[28] It is known as *Vision Zero*, and it has dramatically changed the Swedish approach to road transport system design and operation. The Swedes are not yet at zero, but they are in the top echelon on the safety performance ladder.

The word *zero* has proven troublesome for many nations, some claiming it is unachievable at acceptable cost, and some fearing potential liability claims directed at governments when tragic outcomes occur. Both fears arise because it is governments who design and operate (most of) our roads. Two of the authors of this book were deeply involved in discussions leading to the development of the current Australian national road safety strategy. Unfortunately, proposed titles that included the word zero as an aspirational goal—even the seemingly innocuous "Towards Zero"—were not adopted, and the published strategy glories in the bland title *National Road Safety Strategy 2011–2020*.[29]

The final form of the Australian strategy was achieved by consensus. Some states in their local safety strategies were prepared to go further. The Western Australian minister for road safety, in tabling his state's *Towards Zero—Road Safety Strategy 2009–2020* in the WA parliament (March 2009), provided a strong endorsement of the Safe System approach when he commented:

> Today I ask members to take the moral high ground. I ask members to refuse to accept that death and serious injury are an inevitable result of using our road system; they are certainly not acceptable consequences. Towards Zero is our bold new road safety strategy for Western Australia, and its core theme is that we should never accept road trauma as a fact of life. It will challenge us to strive for zero deaths and serious injuries on our roads.[30]

However, and very importantly, the targets set within the Australian national strategy are for reductions, not in rates as applied in the predecessor strategy, but in the *absolute* numbers of fatalities and serious injuries "by at least 30%" by 2020. It is certainly movement in the right direction, and it is applauded warmly; nevertheless, it is a missed opportunity as the scientific analysis underpinning development of the strategy concluded that a target far in excess of a 30% reduction could in all probability be achieved with current knowledge.[31] The target set cannot be described as aspirational, or even as a stretch target.

Why was the opportunity missed? The answer lies, in large part, in the need to obtain consensus among eight state and territory jurisdictions, each with autonomous responsibility for implementing the majority of the safety initiatives comprising the new national strategy. These jurisdictions have substantially different traffic safety records, and interventions acceptable to one can be anathema to others. We need to break away from jurisdiction versus jurisdiction comparisons and move beyond the futility of a search for "one size fits all" strategies. Some states in Australia are huge in area, with vast regional and remote sections creating special problems in road transport; some are relatively small and highly urbanised. It follows that they will have different safety performances. A similar situation exists in other federations, such as the United States and Canada. If Australia's safety problems had been disaggregated into those experienced by metropolises, large regional cities, regional areas, and remote areas, four substrategies could have been developed—each addressing the differing specific problems. We venture to suggest consensus could have been far more easily reached and a more ambitious national target set. The lack of consensus is more than a lack of political willingness (although the absence of political aspiration is concerning); it is strongly influenced by the invalid comparisons across disparate problems.

This problem is, of course, not unique to Australia; it is shared by all nations comprising federations of provincial jurisdictions that hold much of the accountability for policy and action around road transport—for example, both the United States and Canada.

It is also the nub of the weakness in international benchmarking. Population size is not the only factor influencing the number of disabling injuries and deaths from road transport. Levels of motorisation, vehicle mix, the nature of the infrastructure, and geographical size and topography are all key influencers. The Netherlands, for example, is geographically small (about the same area as Australia's smallest state) and flat, has a relatively low level of motorisation (among motorised nations), and about 40% of travel occurs on motorway standard highways—each of which gives it some natural advantages in achieving a good safety performance. This is not to belittle its excellent traffic safety achievements, which are strongly driven through its adoption of leading-edge public policy and practice, but to underline that our benchmarking should focus on *what nations have been able to achieve and how they have been able to achieve it*, rather than on a blinkered comparison of safety rates.[32]

The most telling argument against relying upon rates to measure safety progress is that it requires us to accept that a finite number of people will receive disabling injuries or be killed each year, and that if either population growth or road use growth is high enough, that absolute number will rise even though our safety rate might be continuously improving.[33]

3.3   WHY DO WE (MOSTLY) RELY ON DEATH COUNTS AND DEATH RATES?

In most countries the police are the initial collectors of data on road crashes and their consequences as they are called to the scene of the most serious crashes and are the first point of contact for aggrieved parties who believe someone else is to blame for their crash.

Not all countries count in the same way something as apparently obvious as a road crash fatality. The WHO standard is to count as a road crash fatality a person who dies within 30 days of the crash in which they were injured, and while most Western motorised countries adopt this definition, many others only include those who die within 24 hours (Indonesia) or 7 days (China) as victims of their crash. Most official records also exclude persons deemed to have died as a result of suicide, or in crashes on private roads, from their tally of road crash deaths. To compare countries, corrections for these differences in reporting standards have to be made.

If this seems complicated, the picture with injury is far worse. There is no agreement among even the states and territories in Australia as to what constitutes a serious injury. There is a move to standardise on a minimum stay in the hospital of 24 hours, although the confounding issue of assessment and treatment delays in emergency departments (as opposed to a stay after admission) is still debated. Some states appear to have a higher total incidence of injury than others, but this turns out to be an artefact of the differences in the legal requirements for reporting a collision. In a state where damage-only collisions must also be reported to police, the minor injuries that often occur, but do not require treatment at a hospital, tend to be included on the report form. Where only injury crashes are required to be reported, it appears that only those where an emergency department is attended get reported; the minor injuries never appear in the records. These differences make comparisons of injury numbers across jurisdictions within a nation, let alone across nations, problematic.

Most Western motorised countries are experiencing decreases over time in the absolute numbers of deaths—albeit that they are relatively small—but there is no commensurate decrease in the numbers of serious injuries. Recent research in the Netherlands, which explored why this might be so, suggests that factors such as lower urban speed limits, more crashworthy cars, and improvements in medical emergency treatment have lowered deaths without reducing total casualty crashes.[34] As we saw in Chapter 2, serious injury can be as devastating in its impacts on not just the victim, but on the immediate circle of family, friends, and colleagues, and if we are to tackle community complacency, we must replace the focus on deaths with a focus on total casualties. A new nongovernment group in Australia adopted the name 33,900[35] to reflect the total casualties in Australia in 2009, and while this is a step forward, we are really no closer to shifting the mindset of the community, and through it the political leaders.

In recognition of the fact that deaths are merely the tip of the road trauma iceberg, there is a major effort underway, internationally, to develop appropriate methods for estimating serious injuries, both within and across national borders.[36] In Australia, while the issue is recognised, we remain hamstrung by a belief among the data gatherers and analysts that we must be able to make definitive counts. Why wait for such a day? Right now we could make reasonable estimates in any of a number of ways. First, we could ask major metropolitan and regional trauma hospitals to provide regular returns as the vast majority of serious trauma cases are transported to such hospitals, and while an underestimate of the true total, this would provide a consistent, current estimate of trends. Second, an analysis could be made of transport injury compensation databases to determine the ratio of serious injury to death from road crashes, and we could use this ratio to ensure the media reported the far larger trauma total rather than being forced to focus only on death. The more the number of deaths is reduced, the less will be the political saliency of traffic safety as a whole, and the harder it will be for the community to demand a heightened level of attention. Some recent exploratory research in Victoria using the transport injury compensation database indicated that while the total number of serious injuries was not falling, the average severity of these injuries was.[37] Nevertheless, at least a quarter of the serious injuries are catastrophic in their severity and long-term impact. We may be making progress, but we remain blindsided by our focus on just the tip of the iceberg.

It is most telling that the UN declaration involves targets set as reductions in the absolute numbers of both deaths and serious injuries, and that the institutional leadership came from the international health agency arising from its forecast that the road toll will become one of the major burdens of global disease over the next couple of decades.

3.4   TRANSPORT SAFETY RATE

The original measure of safety grew from its institutional positioning within the road transport sector of government. In most countries road infrastructure is (overwhelmingly) a public asset provided using taxpayer funds; similarly, traffic management is a government responsibility. While vehicle design as a whole is a matter for industry, vehicle safety specifications are mandated by government regulation, regulation that is driven from the road transport agencies.

Given this institutional setting, it was logical that the traditional measure of safety should be the number of deaths per unit road travel, an apparently simple measure of how safely the road transport system was operating. This rate measure has almost universally been confined to deaths because that is all we have been able to count accurately and in a timely manner, despite it being less than 10% of road trauma.

The institutionally preferred transport safety rate measure is deaths per unit distance driven, for example, deaths per 100 million vehicle miles, as favoured in the United States. (The International Road Traffic Accident Database (IRTAD) reports deaths per billion vehicle kilometres for all countries where it can obtain estimates.[38]) Not all nations have reasonably reliable measures of distance driven, and many nations report deaths per registered motor vehicles (usually per 10,000) as a surrogate. Neither measure takes into account changes in walking or cycling over time, despite the fact that these modes of road use involve higher risks of disabling injury and death; nevertheless, they are the best approximations we have of the level of safety at which the road transport system is functioning, with regard to the inexorable increases in road use over time.

However, it is those inexorable increases in road use that make this measure of safety quite misleading (other than to those prepared to accept that collateral damage is inevitable). Way back in the 1950s, Smeed in the United Kingdom discovered the intimate relationship between level of motorisation (the number of motor vehicles per head of population) and the rate of transport safety.[39] The higher the motorisation level, the better the transport safety performance. Correlation is not necessarily, of course, causation. However, highly motorised countries require better road infrastructure, better traffic management tools, and have populations more experienced with widespread car use. So safety per unit of road use should get better as the system grows and matures. And this is exactly what happens.

The United States is one of very few countries that still use a transport safety measure as its primary measure of progress in traffic safety. One commonly reads, in most official publications out of the United States, that the per mile fatality rate is at its historically lowest point, yet its progress in traffic safety lags far behind that of most Western motorised nations. Table 3.2 compares, over a 40-year span, progress

TABLE 3.2

Deaths and Death Rates over the Past 40 Years

	United States	Australia	Sweden
No. of deaths, 1970	52,627	3,798	1,307
No. of deaths, 2010	32,885	1,352	266
% reduction	38%	64%	80%
Deaths/billion km, 1970	29.5	49.3	35.3
Deaths/billion km, 2010	6.8	6.1	3.2
% reduction	77%	88%	91%
No. of vehicles per 1,000 persons, 2010	841	719	584

Source: IRTAD, *IRTAD Road Safety 2010 Annual Report*, ed. OECD Forum, International Traffic Safety Data and Analysis Group, Paris, France, 2011. IRTAD's data cover the past 40 years.

in the United States, Australia, and Sweden for both the absolute numbers of deaths and deaths per unit of road use.

Several things are worth noting. The motorisation level of the United States is way above that of the comparison countries in this table; indeed, it is far and away the highest in the world.[40] Second, despite the United States having had the best transport safety rate (of these three countries) in 1970, it now has the worst rate, and finally, its drop in the absolute number of deaths over these 40 years is only half that achieved by Sweden and only two-thirds that achieved by Australia. Finally, the percentage drop in absolute deaths in the United States is only half the percentage drop in the deaths per unit travel measure. The system is operating more safely, but still huge numbers of people are killed. Clearly, the fact that the transport safety rate is at its "historically lowest figure" obscures a relative lack of progress in traffic safety in the United States.

The dramatic progress in Sweden may be indicative of a radically different approach to traffic safety, especially when recent progress is examined. In the first decade of this century, Sweden reduced the total number of deaths from road crashes by almost 8% compared with reductions of less than 3% in each of Australia and the United States. Recall that *Vision Zero* was adopted as a philosophy for road transport in the Swedish parliament in the late 1990s. In Chapter 8 we explore in detail how prevailing road use culture and traffic safety performance interact. Equally importantly, we explore how a focus on the rate of transport safety reinforces a belief in the inevitability of death and serious injury as a price that must be paid for personal mobility.

It is important to underline that there is no implied criticism of the approach in the United States in this analysis. The objective is simply to demonstrate that both traffic safety policy and practice in *any* society are intimately linked with the prevailing road use culture. What we seek is a new level of understanding of how road use culture operates to determine progress. Armed with a deeper understanding, each nation will be in a position to make its own public policy decisions in a more informed way.

3.5   PERSONAL SAFETY RATE

In the field of public health, the assessment of the extent of any given disease is its prevalence in the society in question. The debate again arises as to whether the absolute number should be the key prevalence measure or the rate of incidence per unit population, usually expressed per 100,000 persons. The WHO regularly publishes its estimates of the global burden of disease, both to provide benchmarks for comparison and to set priorities for international intervention efforts.[17]

Table 3.3 shows the estimated deaths (from road crashes) per 100,000 population experienced in the various WHO regions of the world in 2008. We can readily see that, on average, low-income and middle-income countries have around twice the road traffic fatality rates of high-income countries.

This rate of death per population is particularly concerning when we realise that the growth in motorisation in many of the low- and middle-income countries is rapid and has much further to progress before reaching saturation. Further, these countries include the majority of the most heavily populated nations on earth (China, India, and Indonesia).

Between 2000 and 2020, in low- and middle-income countries, it is predicted that the number of fatalities will increase by more than 80%, while there is expected to be a further decrease of nearly 30% in high-income countries. Specific estimations by region and the contrast with the aggregate estimate, for high-income countries, are set out in Table 3.4.

The WHO has estimated that by 2030, road traffic injuries will be the fifth highest contributor to cause of death internationally, up from ninth in 2004. Of note is the prediction, in a separate study, that road deaths will be the leading cause of health losses for children (age 5–14) by 2015, and the second cause for men by 2030.[42] Given that younger members of the community are overrepresented in serious road traffic injuries, the future productive time lost due to road traffic injuries, compared with that lost now due to the current highest causes of death—cardiovascular or

TABLE 3.3
Modelled Road Traffic Fatality Rates

WHO Region	High Income	Middle Income	Low Income	Total
		per 100,000 Population		
African	—	32.2	32.3	32.2
The Americas	13.4	17.3	—	15.8
Southeast Asia	—	16.7	16.5	16.6
Eastern Mediterranean	28.5	35.8	27.5	32.2
European	7.9	19.3	12.2	13.4
Western Pacific	7.2	16.9	15.6	15.6
Global	**10.3**	**19.5**	**21.5**	**18.8**

Source: WHO, *Global Status Report on Road Safety: Time for Action*, World Health Organization Department of Violence and Injury Prevention and Disability, Geneva, Switzerland, 2012.

TABLE 3.4
Predicted Changes in Road Traffic Fatalities

World Bank Region	% Change 2000–2020
South Asia	144%
East Asia and Pacific	80%
Middle East and North Africa	68%
Latin America and Caribbean	48%
Europe and Central Asia	18%
Subtotal	**83%**
High-income countries	–28%
Global total	**66%**

Source: Koptis, E., and Cropper, M., *Traffic Fatalities and Economic Growth,* Policy Research Working
Paper 3035, World Bank, Washington, DC, 2003.

heart disease (which usually relate to death in older people)—will be a growing economic impact.

This is a sobering reminder that the weight of the road crash problem not only falls disproportionately on younger people, but on the most vulnerable to disabling injury: riders of motorised two-wheelers and their passengers, pedestrians, and cyclists make up some 50% of road traffic fatalities globally. The proportion of vulnerable road user deaths is higher in low- and middle-income countries than in high-income countries.[1]

The risk exposure of the young is not confined to developing nations. In the United States, road crashes are the single most frequent cause of death for every year of age from the ages of 11 through 27. Further, while deaths from road crashes are the 11th most common cause (the first time out of the top 10 in the United States), they are 5th in terms of years of life lost, which takes into account the age distributions of the various causes of death.[43]

When the measures of years of life lost and disability-adjusted life years are used, there can be no dispute that road crashes comprise one of society's most prevalent diseases—*the disease of mobility*—and constitute one of our most prevalent public health problems. Yet, most nations continue to drive their traffic safety strategies and programs from their road transport-based institutions.

3.6   SO WHAT MEASURE SHOULD WE USE?

Wherever the primary accountability for traffic safety performance lies—typically with a national government, though in some federations it lies at the provincial government level—the short-term targets must be set as specific reductions in the total numbers of disabling injury and death. Eventually eliminating disabling injury and death from road traffic is the only acceptable long-term goal.

Progressing toward zero by reducing the absolute numbers year upon year, despite the inexorable increases in the numbers of road users, registered vehicles, and the volume of road use, is the only acceptable short-term goal. Many motorised nations have reached the point where each new strategy sets a target to reduce the absolute number of deaths below its previous level. In this sense they have adopted an (apparently) acceptable short-term goal. Unfortunately, these targets are typically for reductions far smaller than the scientific evidence suggests is achievable, and hence progress toward the ultimate goal of elimination remains tragically and unnecessarily slow. No nation—other than Sweden—has yet started a public discussion to determine what an acceptable level of trauma might be. A simple "less than now" is not an adequate objective.

Measuring the position of road crash disabling injuries and deaths relative to other public health diseases using age-adjusted measures must also be an important adjunct to targets seeking reductions in absolute numbers. Road trauma is a disease related to our lifestyle, and it is time we conceptualised and treated it as such.

It is also vital that governments set their safety targets quantitatively. Research has shown that nations with quantitative targets in their safety strategies do better than those without, and that those with "stretch" quantitative targets do better still.[44,45]

The Dutch are leading an effort to formulate a composite index of national traffic safety performance to parallel the United Nation's Human Development Index, which attempts to reflect progress toward human quality of life and which includes elements such as life expectancy, average educational level, standard of living, and others.[46,47] This is an exciting step and illustrates that there are visionary leaders in traffic safety trying to shift the field past its antiquated thinking, but there is an enormously long path to travel until the acceptance of the public and, through them, the politicians is achieved.

Providing the traps already discussed are recognised, benchmarking internationally (and within federations and across regions) using conventional measures remains extremely valuable. The objective of benchmarking, which can only happen via the use of rates, is to identify places with low (good) rates, or rapid improvements in their rates over recent history, as an indicator of excellent policy and practice. The focus should then become an examination of the strategies in place in those top performing places and their methods of implementation to draw lessons that might be applicable locally. What did they do, how did they do these things, and how might we apply these things locally are the questions to be answered.[32] Unfortunately, the "best practice" rate is often seized upon as the goal, ignoring the myriad factors that contribute to determining the rate, over and above traffic safety strategies and their implementation.

When benchmarking, both transport safety and personal safety rates are valuable comparators but, as Table 3.2 clearly shows, must also include an examination of the trends in absolute numbers. As nations' road transport systems grow and mature, there is inexorable improvement in both rates, improvement that can obscure a relative lack of improvement compared with what others achieve.

We must, at all times, be conscious of the (unsurprising) tendency by advocates of specific causes to choose measures most favourable to their cause. Let us illustrate. If we examine the frequencies of all causes of deaths in Western society, we find that heart disease and cancer are the "big two." However, as we have already seen, the disease of mobility is a disease of the young. Traffic safety advocates like to discuss "years of productive life lost," being the years from the disabling injury or death event to retirement age. Since cancer and heart disease are skewed toward the older ages, road trauma emerges as a key issue under this measure.

Advocates for the big two disagree and argue for a measure based upon years from event to average life expectancy, which extends far beyond retirement age. These are, sadly, the sorts of games scientists have to play as advocates seeking funds for their field of endeavour. It seems to us that we need to work out how to sidestep unproductive "internal" competition among scientists and ensure society identifies the critical modern public health problems and addresses each.

3.7   HOW MUCH RISK IS TOO MUCH?

Traffic safety is an unintended consequence of road transport, and it behoves us to understand the context in which traffic safety is considered a public policy issue.[48] There are five organising ideas that influence the development of national transport policy:

1. Economic growth is essential for the health of a modern society, and effective transport infrastructure and operation are foundation stones for growth.
2. Social equity is important for the health of society, and universally available access to goods and services is an essential component of equity.
3. Urban amenity (the liveability of a city) is also important for the health of urbanised society, and transport infrastructure and systems impact heavily on urban amenity.
4. Managing the environmental by-products of transport, such as air and noise pollution, and the safety outcomes are important issues.
5. Achieving a sustainable transport system requires a balance between the programs implemented to achieve the immediate objectives and their longer-term ramifications for the future.

The planning of a road transport system available to all in society thus requires an appreciation of myriad concurrent objectives and a capacity to take them all simultaneously into consideration. This is an immensely complex task. It also helps explain why governments set unambitious trauma reduction targets; they make trade-off judgements and, above all else, avoid measures that may, at first blush, prove publicly controversial.

Is it reasonable to seek to set the elimination of disabling injury and death as an overarching objective, to, in effect, become a nonnegotiable given in all planning decisions? *That is exactly what we argue for in this book.* For it to be a reasonable argument, the onus is on us to establish that, in so doing, we can still design and operate a sustainable road transport system that achieves economic growth, social equity, and amenity and avoids environmental damage. And further, that the public can be persuaded to adopt new approaches.

To start with, we need to explore just where personal motorised transport fits in modern Western society, and that is where we now turn our attention.

4 The Car in Society

4.1 CAR DEPENDENCE AND ITS LEGACY

Herbert Hoover's 1928 U.S. presidential campaign slogan "a chicken in every pot and a car in every garage" neatly symbolises the aspirations of "automobility" and serves to mark the emergence of our dependence upon the private car. Globally, one billion cars were manufactured in the twentieth century.[49] In the United States, more than three in every four households had at least one car as early as 1930,[25] and astonishingly, there are now more registered cars than licenced drivers.[49] In contrast, in Australia, as late as the beginning of the 1950s only one in three households had a car. But we then made rapid strides to a point where we now have around 72 cars per 100 people.[49]

As we pointed out in Chapter 1, the car is unquestionably central to the lifestyle of most people in modern Western societies, and the nature of the personal mobility it provides is likely to remain one of mankind's most prized assets for as much of the twenty-first century as we can reasonably foresee.[50,51]

Dorling, in his Westminster oration on traffic safety in 2010, went so far as to say: "We prioritise what is good for the motor industry over what is good for human beings. We do this because the industry has become a force in its own right."[52] Society has such an affinity with cars that Nobel Prize-winning novelist William Faulkner claimed: "The American really loves nothing but his automobile: not his wife his child nor his country nor even his bank-account first."[53] Others have described our car dependency as a Faustian bargain.[51]

There is an intricate relationship between personal mobility, ready accessibility to goods and services, business productivity, and urban development. Urbanisation results from a need for accessibility that, in turn, requires mobility. Mobility values time, which results in supply-side solutions. It has even been said that modernisation is tantamount to mobilisation, and that the car is a metaphor for mobility.[54]

New World countries (such as Australia, the United States, Canada, and New Zealand) have urban forms of lower residential density and greater car dependence. The urban sprawl so typical of these countries results in a need for the private motor vehicle to access essential services such as health and employment, further reinforcing the need for cars as the primary source of personal mobility. Urry comments that the car has essentially reconfigured civil society to incorporate distinct ways of dwelling, travelling, and socialising in, and through, the "automobilised time-space."[55] Old World countries (much of Europe), on the other hand, have higher living densities and, consequently, lower motorisation rates.[56] Table 4.1 illustrates the difference between old- and new-world countries.

In general terms, urbanisation resulted from a need for ready accessibility to essential services as societies became more specialised, which in turn required personal mobility and efficiency in freight transport.[56] The bulk of personal mobility is provided by the car, and the bulk of freight transport is provided by trucks.

TABLE 4.1

Motorisation Rates for Old World and New World Countries

Location	Motorisation Rates per 1,000 Population
European Nations	
France	600 vehicles
Great Britain	565 vehicles
The Netherlands	563 vehicles
Sweden	584 vehicles
New World Nations	
Australia	719 vehicles
Canada (2009 rates)	638 vehicles
New Zealand	734 vehicles
United States	841 vehicles

Source: IRTAD, *IRTAD Road Safety 2011 Annual Report*, ed. OECD, International Traffic Safety Data and Analysis Group, Paris, France, 2012.

Staley and Moore, in a recent book titled *Mobility First*, argue that road transport investment (particularly in the creation of the interstate highway system) brought huge economic benefits to the United States as travel times fell, travel costs fell, freight efficiency improved, and people, businesses, and cities became better connected, but that steadily increasing congestion is rapidly eroding the economic gains and decreasing U.S. global competitiveness.[25]

As mobility is typically interpreted as travel time minimisation, the planning emphasis has traditionally been on supply-side solutions—more and "bigger" roads—while demand management has primarily focussed on getting people out of their cars and onto public transport. While the latter shift has occurred in several large cities, it has only been to a relatively small degree and certainly not commensurate with the total growth in trip numbers. Moreover, fixed-route public transport is not well suited to the increasing frequency of "trip chaining," through which people combine many purposes into a single outing.

Staley and Moore argue that a radical shift is required away from our present system of a "fuel tax funded public good"[25] to a road pricing model where users purchase from private suppliers their access for both route and time of travel, believing that this will lead to private investment in new forms of traffic management and infrastructure provision. If road travel is purchased like any other commodity, they argue, both supply and demand will be optimised through market forces. They do note, in conclusion, that such a radical shift may take decades to achieve despite the technology for automated charging already being a reality!

Rune Elvik, the highly respected Norwegian transport economist, has similarly made a case for the universal introduction of road use pricing. He argues that this will create a true market situation for road transport to replace the dysfunctional quasi-market of benefit-cost analysis.[57]

While toll roads are becoming more common, a widespread public resistance persists toward "privatising" transport infrastructure and its use. The British prime minister, in 2012, rejected tolling for other than new roads, but is reported to be considering "leasing" existing infrastructure to private investors who would reap their return from a share of the vehicle excise duty if performance targets are met.[58] We will see more and more innovative proposals that sneak up on a market-based road use system. Political dances of this kind reflect the need to change our age-old model in an environment where society remains firmly wedded to our present system!

It is also time to question whether mobility really does demand the minimisation of travel time, which is what motorist organisations inevitably clamour for.[59] Much of the motoring public's concern around travel time is not the quantum of time itself, but its unpredictability for a given journey.[25,60] People want to plan departure times efficiently; they want reliability in their forecasts, but as Staley and Moore point out, the "fifteen mile trip from Santa Monica to downtown LA should take about 20 minutes ... but (it can take) an hour or longer—each way." Should we not focus on reliable accessibility—the ability to get safely to where we need to go at a time of our own choosing within an acceptable time frame? Congestion is more a function of interrupted traffic flow than it is of maximum travel speed.

In most nations, proposals for new roads or for upgrades are assessed using cost-benefit analysis with dollar values assigned to congestion reduction (travel time savings), pollution impacts, and safety impacts. In typical analyses, small time savings experienced by huge numbers of drivers add up to a larger total dollar value than does a small increase in deaths and serious injuries. The Swedes, following their formal adoption of a vision of a road transport system in which no one, behaving responsibly, would lose their life or be seriously injured, felt the need to radically change the way they viewed travel time. The Swedes apparently no longer permit the inclusion of travel time savings below a threshold value in calculations of costs and benefits of road proposals. Their mindset is not minimum journey time, but the facilitation of a safe journey per se.[61] (In Section 6.4, when considering the context in which public policy decisions are made, we revisit the role of cost-benefit analysis in more detail.)

Even Staley and Moore, in their passionate defence of the primacy of road travel as an integral part of a competitive modern economy, raise the notion of a threshold journey time that people appear to accept.[25] The fact remains, however, that congestion is measured formally in both Australia and the United States as the difference in the actual journey time between A and B and the theoretical journey time if the entire trip were to be undertaken at the speed limit.[25,62] As anyone with a smattering of mathematics will immediately realise, a trip where the average speed is at the speed limit must contain a considerable portion above the limit to offset those portions (stops at red lights, pauses to allow another vehicle to enter the traffic stream, etc.) below the limit. The congestion measure involves the (very common) assumption—which we explore further in Chapter 9—that the speed limit is not a *limit* at all! Staley and Moore actually propose a range of road design and traffic management innovations that would allow drivers to "breeze along five miles an hour over the speed limit"![25]

The modern mantra, in transport as in many areas of our daily lives, is a search for sustainability. In road transport debates the term has been hijacked by the sectoral interest spin doctors to justify their particular interest:

- Sustainable freight transport means, to both the road freight industry and many governments, transport infrastructure investments and regulatory regimes that facilitate the forecast dramatic growth in road freight volumes.
- Sustainable mobility to governments in Singapore and Hong Kong means keeping people out of cars because of these nations' small geographic areas, high population densities, and concentrated activity centres, while to governments and motorist organisations in the United States it means adding proactively to the road stock.
- A sustainable motoring environment means cleaner fuels, more fuel-efficient cars, stricter emission controls—anything that enables motorists to keep driving unconstrained as before. (A recent survey by a motoring organisation in Victoria revealed that three-quarters of motorists claimed to be worried about the environmental impacts of cars, but they rejected less driving as a solution.)[63]

Sadly, sustainability remains very much in the eye of the beholder. To each vested interest group it means that whenever a balance is sought between what are seen as competing objectives, that balance must favour that group's principal objective! It is time that the objectives of the raft of sectoral interests were made public and transparent and become part of the debate about what we truly want out of our road transport system. As we stated at the end of Chapter 3, our view is that the elimination of disabling injury and death from our road transport system must be a nonnegotiable starting point in all public policy decision making; it must not be permitted to continue to be part of a trade-off process.

4.2   CAR CULTURE

The history of the car is often seen as having two major themes: the car as a symbol of production, harking back to the days of Henry Ford, and the car as a symbol of destruction, capturing ideas of injury, trauma, toxic waste, and pollution.[53] Such simplistic classifications overlook the fact that changing patterns of work, leisure, and lifestyle are wider issues that are tied up in the history of the car, and the strong cultural bonds to car ownership and driving that these have created.[53] This "car culture" is pervasive across all of society. It spans the way in which we develop our cities, the way in which we shape our economy, our views about our "place" in society, and how we function in our day-to-day lives. In many countries it is now considered unusual *not* to drive, with the expectation that we are free to move around our urbanised spaces without restriction.

This culture of cars is also reflected in society's focus on consumption and growth-oriented capitalism.[64] We have an obsession with endless economic growth, requiring continually rising labour hours from households. This places time at an absolute premium, which probably partly drives our relentless demand for high-speed travel.[64]

Paddy Moriarty controversially suggests that if we all worked part-time and reduced our consumption in general, we could all slow down and work far fewer hours![64]

Car dependence is also reflected in broader societal norms. Increasingly, society values "now." The Internet, mobile phones, and social media all facilitate a sense of "nowness," but none are substitutes for the immense personal freedom that the car affords us. Classic movies and books reveal how central the automobile is to our lives, with the "open road" and all its freedoms celebrated in popular culture such as film, television, and music.

Much of the car's role in popular culture is related to the independence it provides.[64] Most New World countries place great value on personal freedom and individualism, and it is this relationship with driving that constitutes an important part of road culture. *Individualism* refers to the belief that the individual is of primary importance, focussing on the virtues of self-reliance and personal independence. The opposing philosophy is collectivism, which focusses more on meeting the needs of the community rather than the individual. Both are socially constructed traits that directly affect attitudes and behaviour, and driving culture is reflected in the extent to which a society places emphasis on individualism or collectivism.

The tension between individualism and collectivism is perhaps best demonstrated through the debate over gun ownership and gun control. In the United States, gun owners often refer to the rhetoric of personal protection and civil rights under their Constitution's Second Amendment, even if collective security may suffer as a result. Much of the individualistic view is that protection is personal—the right to bear arms is to achieve protection—and there is great resistance to institutional means of enforcement. This viewpoint is entrenched within cultural norms, and many sociologists have pointed out that self-reliance and autonomy with limited responsibility to the collective needs of the community is deeply rooted within American culture. As an example, one study found that women in the United States were more likely to view guns as expressions of freedom and independence than women in the UK, who were more likely to view guns as representing violence.[65]

The relationship between individualism, collectivism, and road culture is complex. In countries like the United States, there is a pervasive belief that impinging on a person's rights to drive the way he or she sees fit is akin to living in a "nanny state." This stems from a deep commitment to individual autonomy, which results in a general distrust of government and a resistance to anything perceived as limiting individual rights.[64] In contrast, collectivism is more focussed on sharing resources with others and a reliance on the government for fair and just distribution of collective resources. Countries like Sweden tend to have more of a collectivist approach, and are more accepting of community values and interventions.

Australia is a little schizophrenic—we like to portray ourselves as "larrikins," independent and laconic, but in reality we are very compliant in our acceptance of rules and restrictions.[66] Indeed, our notoriety internationally in traffic safety is the extent to which we depend upon regulation and enforcement. It has been argued that this has stemmed from the combination of a large land mass and a small population that has resulted in one of the lowest ratios of taxpayers per kilometre of public road in any motorised society.[9] Low-cost solutions have long been favoured! A populist distrust of authority coexists with high expectations of government for the provision of services.

Inconsistent government responses to the various problems facing car travel no doubt also feed into the general mistrust of governments. To demonstrate, Paddy Moriarty observes that "during 51 weeks of the year, the [Victorian state] government lectures us about driving carefully: 'speed kills', they scold us. But for the Grand Prix week, 'speed thrills'. Then back to 'speed kills'. When individuals behave this way, we call it schizophrenia."[64]

We know that cars have become part of the fabric of society. Cars are not simply rational objects that we use to go about our daily business; in many ways we have turned them into something of a statement about ourselves and our social identity. The car has not escaped our culture of consumption and commodity. Car ownership itself can be seen as a status symbol,[64] and cars are targeted squarely at what is relevant to popular culture at any given time. Look, for example, at the way cars are named. The ideology of speed and power saw the Cougar, the Jaguar, and the Falcon, and the current obsession with all things technological has welcomed the i30 and the CX7. For young people, cars are "cool," playing into the ideals of style, celebrity and luxury, and having the latest gadget. Thus, the car is not removed from the ideals of society; rather, it is intricately linked with popular culture and idealised with our notion of achievement. Certainly much of the advertising features expensively dressed, attractive models demonstrating how life will be better with the latest and greatest in slick car styling.

Nor is our relationship with cars purely aspirational. Cars are tied up in so much of the way in which we currently live, and the pleasurable activities we take part in; we drive to the beach, the mountains, or the forest, or we go for picnics, all of which create positive cultural associations that form a nostalgic relationship with the car. In recent years, there has also been an increasing association with the car as "caring parenting"—from using the car to soothe an infant to sleep, to protect children from the elements on the way to school, and the idea that one stage of parenting is akin to the role of a chauffeur by taking the children to all their activities.[53]

In contrast, the academic flavour of research in traffic safety is neutral, which serves to reinforce the notion that safety is an issue for potential victims, not for the car and its driver, and so the problem becomes one of perception and balance of risk.[53] This essentially means that our sense of experience around the daily use of the car, and its relationship with family life, is hard to extricate from the more destructive aspects of the motor vehicle.

Cars are incredibly central to modern life and popular culture. Think about all the car chases in the movies or on television shows, the car's role in teenage dating movies, and even television shows dedicated not to decorating homes, but to decorating cars (*Pimp My Ride* and *Monster Garage*, to name but two). In fact, cars have been so entrenched in our culture that we have personified, or anthropomorphised, them in movies and cartoons. Classic examples are Herbie (the *Love Bug*), KITT from the *Knight Rider* series, and Lightning McQueen from the movie *Cars*. In many ways we have given these cars features we would like to see in ourselves, and so we lose the idea that the car is actually a functional object (Figure 4.1).

This point is reinforced by research on car preferences. Studies have found that we prefer curvy cars, which some researchers think may actually have links to our evolutionary past. Humans are hardwired to see faces in everything, which would have likely protected our ancestors—mistaking a stone for a bear is not so much of an issue, but the reverse could prove fatal![67] One study in Austria looked at the tendency to humanise cars, and found that people attribute human traits to vehicles based on features such as the shape of the headlights and the size of the windshield. The researchers thought that it makes sense that people personify cars; the fronts of cars are symmetrical and have features akin to people, such as "eyes" represented by headlights and "ears" denoted by side mirrors.[68]

Similarly, a study at the University of Vienna presented a group of participants with pictures of cars. They found that one-third of the participants associated a human or animal face with at least 90% of the images presented.[69] Many of the movies that have humanised cars have extensively mined these features to create characters that viewers can identify with. These studies highlight how we apply a range of human or animal traits to cars, which is a process not often seen in functional objects or material goods—we are much more likely to name our car rather than our cell phone, for example!

Daniel Miller gives a fascinating and slightly different account of how we construe the humanity of the car. He gives the example that a human foot could easily be analysed as a technical construction of bone and flesh that provides an individual with a means of mobility. On the other hand, the foot could equally be seen as integral to the individual's humanity; bipedal locomotion is what distinguishes us from so many other species. Miller argues that we are so socialised as to take mobility for granted, we think of our world through a sense of self in which driving, roads, and

FIGURE 4.1 Example of friendly features ascribed to a cartoon car. (From iStock.)

traffic are simply integral to who we are and what we presume to do each day. This makes it quite understandable that we see cars as being very much a part of our sense of self and not just objects to get us places.[53]

On a different note, an interesting finding in the University of Vienna study was that both males and females liked "angry-faced" cars. This idea of aggressive-looking cars feeds into another significant cultural issue around motorisation—the *masculinity* of cars. The notion of gender is constructed from a range of cultural and subjective meanings that change according to time and place.[70] The idea of masculinity is well documented; we know that males experience great social pressure to endorse a range of gendered societal prescriptions.[70] Take, for example, health-related beliefs, such as men are independent, self-reliant, strong, robust, and tough. These beliefs have been shown to lead to all sorts of problems for men, such as delayed treatment for illness and poor health outcomes.[70]

The car in today's society is typically associated with masculinity. This myth is exacerbated by the aggressive advertising of cars, which targets themes of masculinity to generate sales, for example, power, performance, speed, strength, and status. It is not often, though more frequent than ever before, that you see safety features as a selling point! The car as a masculine status symbol plays out a multitude of unsafe behaviours in advertising, for example, rally-like driving, drag racing, or going from nought to 60 in however many seconds. Although many countries have tried to put a stop to these aggressive advertising techniques, the impressions are lasting and many of the *underlying* principles are still the same—just packaged in a different way.

The masculinity of cars essentially plays into the ideals of power and status, aspects that are also socially constructed. It raises the possibility that people see dominant or masculine-looking cars as a benefit in daily "battles" with traffic, potentially playing a role in road rage.[69] It also talks to the notion of boys "misbehaving" in cars, an issue that has been culturally entrenched for decades. Generally, the phenomena of dangerous driving and "cruising" have been predominantly seen in males—boys and their toys.[71] However, it is possible that we will see a shift in the gender distribution in the coming years. Researchers are talking about the adoption of a "ladette" culture, which is signified by a shift in female culture to "women behaving badly," or adopting masculine behaviours—such as binge drinking and violence—almost in competition with men.[72] A host of television shows like *Girls Gone Wild* reinforce this closing gender gap from little-known subcultures into the mainstream. This type of culture shift indicates that many behaviours that have typically been associated with young males are now being seen in young females. A case in point is "hooning," which has been almost restricted to young men in the past, but is now being reported increasingly among young women.[73] What this all tells us is that we can't ignore the deeply rooted cultural associations we have with cars that continue to evolve as society changes.

4.3   DRIVING CULTURE—RIGHT VERSUS RESPONSIBILITY

We have seen that the car itself is integral to modern society and popular culture, and that we have developed a unique relationship with the car. Underlying all of this is the actual experience of driving. There are many aspects to driving, including individual behaviour, social expectations, and cultural norms for the traffic environment (e.g., do we drive on the left or the right? may we turn left on a red light?). In Australian capital cities, about half of all car trips are less than 5 kilometres, a distance most people could cycle without raising a sweat. However, social norms and a strong focus on people having cars mean that most people will drive for these trips, mainly because it is the "done thing." This suggests that our experience of driving is not only what we *want to do*, but also what we are *expected to do*.

One aspect of the driving experience that we hear a lot about is speed, a theme that has long been celebrated as indicative of the "thrill of an era." However, Daniel Miller raises an excellent point about our subjective, and lived, experience of speed. When speed initially moved into mainstream culture (around the 1930s), the experience of speed was travelling on the open road, in an open-topped car, at speeds of around 30 mph (at a push). At the time, this would have felt like an authentic speeding experience. However, when you take speed into the modern context and account for all the current technology, travelling at 800 km/h on a plane feels painfully slow (although, presumably it wouldn't feel that way if the plane decided to nosedive). Similarly, travelling in a contemporary, sealed car with good suspension at 130 km/h doesn't feel that fast. And that is precisely why it is so dangerous.[53] So although speed is a deeply celebrated part of our culture, as we have chased the experience of ever-increasing speeds, we have also failed to account for the very serious consequences high-speed mobility has.

Another aspect of the driving experience that we focus heavily on is that what we *actually do* inside our cars makes them so important to us. We conduct much of our lives in our cars; they are like mini worlds that are an extension of our private space, very similar to our homes. The Virginia Tech 100-car study provides a fascinating account of all the things we do in our cars[74]; we eat, we drink, we listen to music (of our choosing), we sing, we tap our fingers on the steering wheel to the beat of the music, and we throw tantrums at other drivers who *clearly* can't drive as well as us. Often the car is the only place that we can be alone. It provides us with security and protection from the weather, makes it easier to move things around (such as children or shopping), and provides a means of mobility for the elderly.[53]

Critically, cars also provide us with privacy. When we are inside our cars, we see this as private space where we have the *right* to do as we please. In current society, we place a great deal of emphasis on *rights and responsibilities*. For example, if we need to go to hospital, we are duly informed of our rights and responsibilities and asked to sign an agreement. There are fancy brochures to take home and digest at our leisure, to ensure that we recognise that our health is a partnership approach and that this balance between our rights (e.g., to privacy) and our responsibilities (to ourselves and our health professionals) is the key to success. Yet we don't view cars in the same way. Rajan comments that the problem is that the car has become much more of a right than a responsibility; thus, this balance of power is lost: "to drive and operate automobiles has become almost the inalienable right of every individual to achieve goals and purposes efficaciously across a space specifically engineered for this purpose."[75]

Thus, the balance of rights and responsibilities for an individual driver is obscured by fundamental questions about citizenship and contemporary civil society. Through our steadfast belief that we have the right to be on the road, we often forget that driving is, in reality, a social behaviour conducted in a public place. We permit *licenced* drivers to operate *registered* vehicles on *public* roads under a set of *regulated* conditions. Driving is, in objective reality, a *responsibility*, not a right. The authorities, at least, expect that driving behaviour will be socially responsible, predictable by fellow road users, and compliant with formal system rules.

There is ample evidence to show that while we know what is socially expected—and while we typically give socially desirable responses to surveys—we (as a collective of road users) frequently do not let this knowledge determine our personal behaviour. Psychologists distinguish between declared preference (what we say we do) and revealed preference (what we actually do). There is a social incentive to give an acceptable answer! For example, recent surveys of motorists in the United States[59] found that

- More than half said texting/emailing while driving is very dangerous, but one in four admitted sending a message and one in three admitted reading an incoming message while driving in the past month.
- 45% admitted speeding more than 15 mph over the limit on a freeway in the past month, and one in three said it's OK to do so.
- Speeding in residential streets was considered one of the most dangerous behaviours, but 25% admitted doing so in the past month.
- Oddly, in light of the above two responses, two in three favoured increased speed enforcement!
- One in three admitted to going through a red light in the past month.
- 48% opposed a proposal for a 10 cent per gallon levy for dedicated expenditure to make roads safer.

Similar results have been reported in Australia and New Zealand.[63]

What can we discern from such survey response patterns? Cell phone use, speeding, and red light running are all behaviours with immediate personal rewards, and these responses indicate that such rewards take precedence over the alternative behaviours, which would contribute to community safety outcomes. At a personal level we are not willing to sacrifice immediate personal gains for the greater good. (And, of course, the contribution to traffic safety is, at best, indirect and manifest only when the vast majority of people engage in the socially compliant behaviours.)

It is clear that we place a great deal of value on mobility as a society. It is also clear that our commitment to unrestricted driving is at odds with the more destructive aspects of the car, such as trauma or the environmental impact. For example, the response to the survey question concerning a possible fuel levy is revealing—the immediate personal disincentive of higher costs at the fuel pump serves a psychologically identical purpose to the immediate personal gain of, say, speeding to catch a green light! While 75% of Victorian motorists are worried about the environmental impacts of the car, they do not see less driving as a solution. Instead, there is a demand for cleaner fuels, more fuel-efficient cars, stricter emissions control—anything that enables us to keep driving as before![63]

4.4   RISK-TAKING BEHAVIOUR

We accept risk at a personal level because we believe that risk management is entirely within our personal control. It is this sense of personal control that cements a belief that most victims of road crashes are not victims at all, but are to blame for the crashes in which they are injured. Jiggins, in his analysis of the way crashes are portrayed in the media, talks about a "villain alongside a victim."[76]

"Blame the victim" is not unique to traffic safety. In the United States, where lung cancer claims more lives than breast, prostate, and colon cancers combined, it has been argued that lung cancer cannot get public traction because it is seen as self-inflicted! As a consequence, it is uniformly underfunded (underfunded and underresearched based on the amount of research dollars allocated per death).[77] The Canadian Cancer Society has implicated the lack of research funding as a factor for the generally poor prognosis in lung cancer (the incidence of lung cancer is not higher than other cancers; however, it kills more people, more quickly). The perceptions of fault are widespread, and can even be found among health professionals.

The most frequently heard solution, chanted as a mantra, involves some form of education and training. The call is to improve the skills of all the people that can't possibly drive *nearly* as well as us, and once people are upskilled, no more crashes will happen. This never accounts for human error, near misses, commonplace mistakes, or sudden changes in the road environment (some gravel on the road or an animal). The enthusiasts will tell you that it is their level of driving skill that ensures that they match their driving to the conditions, and they know how to handle the car in any situation. The question is, though, what happens to these superior driving skills when we hit an unexpected pothole or a child runs out in front of the car? When we get into our cars each day, what we are *actually* doing is strapping ourselves to 2 tons of glass and steel without *ever* being able to truly predict the road environment, no matter how good a driver we are.

The misplaced belief in education as a panacea is not confined to traffic safety. Levitt and Dubner use an example from medicine: "In the modern world, we tend to believe that dangerous behaviours are best solved by education."[12] Doctors wash their hands on fewer than half the instances that they should, despite hospital policies dictating hygiene standards. Of course, it is not the doctor's life at risk, and he or she gains precious time by not complying. We are indeed driven by personal incentives! Decades of public education to promote seat belt wearing failed to raise (voluntary) wearing rates above about 40% in Australia.[78]

Much has been written about risk compensation—the "dance of the risk thermostats."[79,80] Man does not seek zero risk. Society places a premium on risk taking in many aspects of our lives, especially in business and in sport. The risk homeostasis theory posits that man operates at a level of personally acceptable risk, and that if interventions are applied to reduce risk in a given situation, individuals will change their behaviour to restore their personal risk to its comfortable level. For example, if seat belt wearing is seen to reduce the risk of injury, a driver will likely increase his or her travel speed. Taken to its extreme, this would imply that we are wasting our time in redesigning our road, vehicle, and traffic management systems in ways that reduce risk. The evidence is unconvincing. While it seems that in some instances we may trade a perceived decrease in situational risk with more risky behaviour, it seems we trade only a small portion of that change in situational risk.

We perceive the risk of a serious casualty crash, at the personal level, to be very small. This makes perfect sense—we drive for almost every conceivable activity as a part of our daily functioning, and our personal experience of danger during this daily road use is very low. On any given individual trip, the risk of a crash (let alone a serious injury) is indeed miniscule. A traffic safety specialist has described the risk to an individual as driving the equivalent of 30 round-trips to the moon before being killed.[13] The economists Levitt and Dubner, in their book *Superfreakanomics*, estimated that "if you drove 24 hours a day at 30 miles per hour, you could expect to die in a car accident only after driving for 285 straight years."[12] Such "averages" are, of course, as misleading as saying the average human being has one breast and one testicle! What we fail to appreciate is that the aggregate result across society of millions of low-risk trips is a large absolute number of disabling injuries and deaths. Why should we view the risk only at the personal level? Why are we not concerned about the overall effect on our society?

Media reporting of road trauma doesn't help matters, as only deaths are routinely reported, and deaths are merely the tip of the iceberg. There are well over 10 serious injuries for every death, and injuries, as we saw in Chapter 2, have impacts well beyond the injured person. Occasionally we read of the financial burden of the road toll, but this number—currently in the billions of dollars per annum in Australia—does not resonate with the public, as lobby groups and politicians are continuously quoting "big numbers" for every cause they want to promote, and most folk have little concept of what billions means.

In effect, we drive the way we do because we see it as our right, not because we have some notion about a communal responsibility. Why, then, do most people comply with laws and road rules? For the most part, it is because it's in our self-interest to do so; motorists are deterred through fear of detection and the penalties that they would incur otherwise. There is also a strong element in most Western societies of obeying the laws because it is the *right thing to do*, much less so that people think it is the *safe thing to do.*

To demonstrate, general adherence to road rules seems to depend on which law we are talking about. Some people may speed at 2 a.m., but will still stop at red lights. Part of this is that there is limited respect for speed cameras because they are seen as a form of government corruption—they are perceived to be about revenue raising, not about safety. There is, as yet, no widespread acceptance of the science that shows conclusively that small changes in mean speeds produce large changes in crash risks and larger changes in fatal crash risks. (We explore the science and the culture around speeding in detail in Chapter 9.) For some reason, scientists and governments have never been able to adequately convince the public of the facts. This is particularly relevant because all of our incentives are either pro-speed or antisafety; we get there faster, we make short-term gains that are tangible to us at the time.

Ensuring that people understand *why* the road rules are in effect is as important as knowing what the consequences of breaching them are!

It is also important to stress that many of the behaviours we see while driving (particularly unwanted ones) are closely related to system design and operation. Tom Vanderbilt comments that the way we drive is essentially a product of the road environment, not of human behaviour itself![60] Expressing the same point from the other end, David Shinar wrote: "Understanding driver behaviour, driver sensory, cognitive and motor skills and limitations, driver motives and attitudes is a key to improving highway safety. But like a New York City apartment, it takes more than one key to unlock the door.... The role of vehicle and environmental improvements in this context is to compensate for drivers' shortcomings."[81]

This research tells us that we drive the way we do for a multitude of reasons, but on the whole, traffic safety doesn't rate particularly high on our personal radar. This is challenging for safety professionals, given that most traffic safety activities are focussed on changing individual behaviours, and not on changing the way in which we view traffic safety more broadly!

4.5 PREVAILING CULTURE OF BLAME

It is clear that we have a complex and multidimensional relationship with cars, and we see cars as far more than objects with specific functionality. The car is often used as a status symbol, a way of demonstrating where we fit in society. In fact, some companies even ask to see your car as part of the interview process. The theory is that the way you keep your car (and by default, what you drive) indicates what kind of person you are, and how you conduct yourself.[82]

So, we believe that cars speak volumes about who we are, but how have we arrived at this way of thinking, and where is this information coming from? Print media, television, and radio have long been the primary sources of current affairs information for the general population. Each of these forms of media, along with the latest advances in information sharing, poses its own challenges for the traffic safety cause. To understand these challenges, it is important to first consider the context of media reporting.

What happens when a journalist decides to cover a road crash? Usually, the reporters are sitting at a desk listening to a police scanner, waiting to decide which of the unfolding events are likely to grab public attention the most. Print media in particular needs to contain a sensationalist or very dramatic angle in order to do so, and so the ultimate consideration becomes how much suspense can be extracted from the event, and whether the quotes and pictures are likely to be dramatic enough. Occasionally serious injury crashes are reported, but they are generally not followed up unless the person dies. So, the initial crash is sensationally reported, but that 30-year-old guy who became disabled in the crash is never (publicly) seen again. The longer-term consequences are rarely followed up to see what his life is like now that he can no longer walk, talk, or feed himself, or pick up his children.

Television media has the added problem that it can generally only report things where there is dramatic footage to drive the story. Given that crash scenes are sanitised very quickly (at least in Australia), there are usually no graphic images to run, and journalists would need to be essentially acting in real time to get good images of a crash. This problem is exacerbated because often the images that are supplied to the media after the fact are when the vehicle was badly damaged (with the implication of high speed), or where controversial imagery can be seen (such as bottles of alcohol). This reinforces the perception that serious crashes only happen when someone is doing something reprehensible.

How does this play out in a typical story? We generally hear "breaking" news only. There may be a frenzied report about a crash that involved high speed, alcohol or drugs, or five young people carried in the luggage compartment that shows a dramatic scene swarming with police and ambulances and flashing lights and distressed witnesses and lots of mess on the road. Early "reports" suggest that it looks like speed *may* have been a factor, but this can't be determined at the time and it will take the police some time to investigate. By the time the investigation of the crash has been completed, the media interest is long over and the full panoply of causes and consequences never comes to light. Unless of course there were actually five people in the luggage compartment, in which case everyone wants to hear about that! In the current media climate, reporters act on the best information available at the time, and retractions or corrections are not necessary if it turns out the details were not quite correct. So the readers or viewers are left with the lasting impression that, yet again, it was another speedster who was up to no good and got his "just desserts."

Journalists are the gatekeepers of information that is essentially sold to the public as truth, a level of power that is often not adequately recognised. In order to function well in their jobs, journalists need to give people what they want and capture the attention of the public in a matter of seconds, all while under enormous workload and time pressure. In the current technological climate, we have seen a shift to online media, and by moving to a culture of immediacy, the face of journalism is changing. News is provided in real time, fed by the expectation that images and stories are accessible as the event is unfolding. Long gone are the days of carefully researched and thought-out articles. In some situations, there appears to be a culture of getting the story at all costs—the recent *News of the World* telephone-tapping debacle is a classic example.

Some of these issues were explored in recent research by the Centre for Advanced Journalism at the University of Melbourne. The centre conducted a study about the way in which the media covered, and responded to, Victoria's 2009 bushfires.[83] Dubbed Black Saturday, the bushfires represent the worst peacetime disaster in Australian history, claiming 173 lives, injuring more than 5,000 people, killing countless animals, destroying thousands of homes, and burning more than 4,500 square kilometres of land. The media coverage was intense, and the results of this research, which interviewed media personnel about the events that unfolded that day, are controversial.

Two main issues emerged: first, attempts by journalists to circumvent roadblocks set up by the emergency services to control access to the affected areas; and second, the use of deception to gain access to fire sites by some media personnel.[84] This suggests a significant lack of ethical standards among some journalists, with a higher value being placed on successfully meeting competitive pressures to inform the public than on their ethical duty to respect the law.[84] However, it is also telling that none of the journalists included in the research had received any kind of briefing, nor were they told what to expect when covering the bushfires; most stated that they knew what they had to do, even if it meant going against their conscience or better judgement. In essence, no boundaries were set, and it was left up to the individual's personal moral compass.[31]

What emerges from this research is that there are no agreed principles on which the media and authorities operate following a disaster, or any other type of trauma, and there are insufficient consensual ethical standards among journalists. It would seem that individual journalists make up the rules as they go along, which leaves the door wide open for errors in judgement, inconsistencies, blind eyes being turned, ethical lapses, and compromises being made.[83] The authors concluded that an "ethical free-for-all" developed in the aftermath of Black Saturday, with some media personnel (and indeed, emergency services personnel) making "exemplary ethical decisions," and others making extraordinarily poor ones.[83] This approach is entirely unsatisfactory from every point of view: the media, the authorities, the survivors, and their communities. Similar themes can be observed over and over again when it comes to the reporting of road trauma.

It is important to stress that we are not engaged in media bashing. At the end of the day, media conglomerates are businesses that focus on their bottom lines. Many journalists do an exceptional job under extremely difficult circumstances, and mostly, they are simply reporting the best information they have at the time. It is also important to realise that there are great opportunities for the media to contribute in the public health and public policy arenas. Certainly, part of the way in which the community views traffic safety is shaped by reporting in the media. But the public also has to take responsibility for what it wants to consume; while we demand to know about hoons and speed, then that is what sells and that is what will be reported. When it comes to traffic safety, maybe it is just too close to home, so we prefer to read about what happens to "other people," the ones that can't or won't drive responsibly, or are totally inexperienced or are hoons with five people in the luggage compartment.

We see the same findings over and again in content analyses of media reporting of road crashes across a wide range of countries.

Dann and Fry analysed media reports of road safety issues between 1996 and 2007 in Australia.[85] They found the coverage increased almost 15-fold from an average of 2.6 mentions per month in 1996 to 1.2 per day in 2007. The most telling statistic, however, is that the vast majority of the mentions were negatively framed—they focussed on horror, stupidity, tragedy, and victim blaming—and that this negative framing increased from 84% in 1996 to 99% in 2007. The mentions were largely about "sensationalising individual trauma." The authors concluded that, since media coverage is the opposite of the overall gains that have been made, there is, for the community, a perceptual failure of road safety efforts. Failure is not only visible, but a visual spectacle via a crash, whereas success is an invisible benefit—things that have not happened. They suggest that road safety is not a value-neutral proposition, and this helps explain why the community opposes resource expenditure or further restrictions on freedoms.

Martin and colleagues report a 2004 U.S. study that examined 846 fatal crashes in an official federal database. Only one-third of the crashes in the official record had been reported in the press (interesting in itself), and among those reported, teen drivers and drunk drivers were overrepresented, with most stories attributing blame to a villain.[86] Heng and Vasu analysed newspaper reports of crashes in Singapore for all of calendar year 2007.[87] Almost three-quarters of the reports assigned blame to one or more of the involved road users. Notably, there were positive mentions of enforcement in over 70% of reports, but little about injury prevention efforts or their effects. Clearly, we are missing opportunities to raise community understanding.

Graham and White state: "Sensationalist reports in the current affairs media may give the impression of 'hoons taking over our streets' and of P-platers being inherently dangerous drivers."[71] The reality is that while young, novice drivers are markedly overrepresented in casualty crashes, the majority do not crash, and hoon behaviours such as street racing, burnouts, doughnuts, and the like account for a very small part of total trauma. Recent research at the University of Adelaide examined in detail 256 nonfatal crashes involving drivers aged 16 to 24 and concluded: "the majority (70%) of young driver behaviour leading to crashes was not caused by risk-taking but due to … simple errors in which they failed to use routine safe operating practices."[88]

MacRitchie and Seedat analysed media articles on road safety during Easter and Christmas seasons in South Africa in the mid-2000s. The common theme in reports was that the roads became a "war zone" during festive periods, with carnage dominated by irresponsible speeders.[89] Davison concluded that public perception around traffic safety transcends the utilitarian logic of prevention and cure and challenges the very values of freedom and material progress that inspired mass motorisation.[51]

While the nature of media reporting reinforces the blame the victim belief, we cannot lay the blame at its door; reporters are mirroring input from many sources. The police investigate serious crashes in order to press charges against illegal or reprehensible behaviour, and the insurance companies vigorously seek ways to limit the size of their payouts. David Shinar quotes the highly accurate but superbly ironical Stannard's law: "Drivers explain their crashes by reporting those circumstances of lowest culpability that are compatible with credibility!"[90]

The culture of blame is at the centre of a vicious circle. Dare we suggest that governments do little to challenge the belief either, as it keeps the pressure off the level of expenditure that would be required to build and operate a truly safe road and traffic management system.

4.6 VESTED INTERESTS AND THE RISE OF "ANECDATA"

We are witnessing a huge shift in our way of accessing information; we are no longer reliant on mainstream print and television media. Information is immediate and relentless; websites abound providing an unfiltered communication environment. Anyone can post whatever information they wish, and people can always find material to suit their existing view of the world. Truth has become an issue of convenience as the Internet has shrunk the gap between the expert and the public.

The combination of the sheer volume of material immediately available and the total absence of quality control has adversely affected the level of debate around issues of public policy. As Australia's former minister for science said in 2011, when comparing political debates in U.S. presidential campaigns a century apart: "In 1860 the technology was primitive but the ideas were profound and sophisticated. In 2011 the technology is sophisticated but the ideas ... are embarrassing in their banality, ignorance and naivety."[91] The same can be said for debates at the political level around such issues as speed enforcement in Australia, Canada, and the United Kingdom (see Chapter 9).

There is a balance to be struck in the use of social media to inform the public during an unfolding crisis or trauma situation. Real-time communication is now considered the way forward by providing a cost-effective and efficient means of keeping the public informed about developing situations. However, we also have a situation where every car tends to contain at least one mobile device with a camera or video that can immediately upload images and videos into cyberspace. Many countries have a level of censorship on printing pictures of cars (and victims) after serious crashes, to protect both the victim and their families, but this censorship is almost impossible to control with the widespread use of mobile devices. This has the potential for further desensitising us to the effects (and aftermath) of a serious crash; we see the crushed metal and bit parts of the cars, but not what happens in the hours, days, or years that follow.

Adding to this overall problem is the recent trend in memorial sites on social networking outlets. These create significant exposure to road trauma, and place the public outpouring of grief front and centre for more mainstream forms of media. Theoretically, this could increase exposure to road trauma and provide a springboard for action, but it seems to be having the opposite effect. Once again, it is young people that tend to mourn their losses in this way, so by the time it hits mainstream media, we are actually being further exposed to the tragedy of young people "behaving badly."

There is a second potential problem. We know that social media in all its forms is expanding at an incredible rate, from general social networking sites to vested interest websites and all their associated forums and message boards. This has opened up the world to direct communication with anyone, anywhere, at any time. Academic research tells us that people tend to make decisions and form their opinions using networks of peers. However, no longer are decisions being made based on discussions with a few close peers; instead, there is a whole world of opinion to choose from. The biggest problem with this is that we know nothing at all about the background or vested interests of a random, faceless individual on the Internet, even if we feel we know him or her well through interactions on forums or message boards. There is no way to tell whether these opinions are informed, and people can present their opinions (or anecdotes) as "fact" entirely unchecked. This phenomenon is so widespread it has led to the birth of the term *anecdata*, which is the use of "multiple points of anecdotal data to confirm any stipulation or compilation of correlated stories or other single pieces of information produced to appear like actual scientific data."[92]

Of course, social media are not all bad; great support can be found for traffic safety. For example, significant discussion can be found on parenting forums about the correct and safe use of child restraints, how to teach young children good road sense, or how much driving supervision is needed for teenagers. Social media also provide a platform for traffic safety groups like the U.S. Mothers Against Drunk Driving (MADD) to get their message out without the significant costs associated with advertising. The more challenging aspect is how to spread these traffic safety messages beyond their microcommunities to a much broader, society level.

Not only are vested interests making their way into traffic safety through social media, but the role of lobby and interest groups is also an important one in the disconnect between car culture (or car rights) and traffic safety. A clear example is all the TV shows, magazines, reviews, and blogs geared toward "motoring enthusiasts." The basic premise is to "test drive" cars, usually at high speed, which sends the message that this is the type of driving and performance that we are actually testing the cars for. This conveniently forgets, of course, that these tests are performed in controlled environments, usually with protective gear (and no other traffic or obstacles!), and bears no resemblance to driving in normal traffic on normal roads.

Having such focus on power and performance is problematic on many levels. What we really need to do is make a clear distinction between driving as sport and regular driving in our daily lives that gets us from point A to point B. Using cars for sport is fine in the right context; there are dedicated tracks and meets and races to ensure motor sport can be played as safely as possible in an environment where harm to the drivers, and others, is minimised. However, the boundaries have been blurred. This type of driving has absolutely no place on normal roads, and yet we continue to feed this myth through the widespread consumption of motoring "enthusiasm."

The lawyers have long benefitted from civil litigation arising out of road crashes, born from the belief in misbehaviour and negligence. They fought the introduction of a per se blood alcohol law because it curtailed their defences of clients charged with drunk driving. They then turned to challenges of the equipment until case law established the reliability of the measuring devices.

The introduction of Victoria's no-fault transport injury compensation system dramatically reduced civil litigation. In recent years, however, there has been an increase in common-law claims as lawyers have started advertising their no-win, no-fee campaigns, with information on their websites such as "have you received all you are entitled to?"

Interesting commentary can also be found on the Internet in forums and chat rooms in response to any new research or policy that advocates a safety approach. Many of these motoring enthusiasts vehemently criticise what they perceive to be the anticar culture, which they believe impinges on their rights to drive in the way that they choose. There are some fascinating websites. SENSE (Safety by Education Not Speed Enforcement) rages against attempts to slow traffic, arguing that each driver should be free to choose the speed at which he or she feels appropriate to the conditions. Such sites should be a source of amusement, but sadly, material on the Internet is widely accepted automatically as "truth," especially when it supports one's own view of the world. Mind you, experts often disagree, muddying the waters considerably.

4.7 CELEBRITY CULTURE

Increasingly, we are living in a culture of celebrity. The obsessive and relentless following of celebrities is manifested throughout popular culture; there are even websites dedicated to what they do on a daily basis, what they are wearing, and where they are at any given time. This can serve some really useful functions; for example, many celebrities promote great causes like breast cancer, or People for the Ethical Treatment of Animals. Younger generations look up to the latest celebrity (or sportsperson) as a role model, and celebrity affiliation with these complex problems has brought some really challenging issues front and centre into public debate. Prostate and breast cancer are great examples—just normalising the words *prostate* and *breast* and encouraging discussion is half the battle.

Unfortunately, there seems to have been an opposite effect for traffic safety. The culture of celebrity (and associated wealth) has somehow transitioned into fast cars and, it would appear, in many cases living above the law. For example, take the number of drug and gun-related convictions among celebrities, and in particular, look at the number of high-profile driving convictions. Driving while under the influence of alcohol or drugs has been reported right across the celebrity spectrum, from newbies to old-timers, and magazines gleefully print the latest celebrity mugshot seemingly every week. This tells a very clear story that those celebrities that so many aspire to be like really don't think there is anything wrong with a bit of drunk driving. And generally, they are "let off" anyway.

There have also been several high-profile hooning incidents in recent years. For example, Lewis Hamilton, the Formula One racing driver, had his car impounded for show-off speeding on local roads following the Melbourne Grand Prix. Although this case should act as a deterrent, what it seems to have done is normalise this type of behaviour. The public debate didn't focus on safety and what to do about hooning behaviour; rather, it ended up focussing on whether, and why, the police "target" specific people.[93]

4.8 OVERALL CULTURAL CONTEXT

The community is assailed with mixed messages about cars and driving from many different angles. We hear about the luxury, power, and performance of cars from advertising, the requirement for unfettered mobility and economic sustainability from our governments, the polluting effects from the environmental authorities, and the consequences of breaking the rules from the police. The most constant message is that driving is under our control, and that only miscreants and incompetents have crashes. Even the laudable public education traffic safety efforts of government and nongovernment bodies tend to focus on specific high-risk behaviours. Listening to so many ideas and competing agendas can have a detrimental effect by putting safety in the "too hard" basket and causing the bulk of the community to disengage from the entire debate.

A parallel disengaged context arguably exists around the case for policy and action with respect to climate change. Generally speaking, the scientific debate focusses on the mechanism by which climate change is occurring, and how rapidly that is happening, not on whether it is happening *at all*. However, the disagreement among the scientific community creates the perfect platform for climate change deniers to base their argument. What is actually being communicated to the public is that even the scientists can't agree on what is happening; therefore, it can't be that much of a problem, especially since almost all suggested actions involve some constraints or sacrifices. The experts argue their case, in language that the majority of people do not really understand, and the press picks up these sound bites, which are played over and over.

Compounding the problem, the existence of public controversy plays into the hands of the media who facilitate uninformed discussion by giving more than equal time to nonexpert opinion. A British author wrote in an opinion piece in the *Guardian* newspaper: "The press is further littered with climate 'heretics', almost all of whom have academic backgrounds in history, literature and the classics, with a diploma in media studies. (All these examples are true)." He opened his article with the rhetorical question: "No one would want a novelist to perform brain surgery with their ballpoint pen."[94] In the same way an antispeed camera group in the United Kingdom used an academic with a PhD in sociology to evaluate the effect of cameras on casualty crashes.[95] His analysis was demonstrably inadequate scientifically, but the damage was done as the resultant scientific debate contributed to public confusion.[96]

When it comes to traffic safety, one of the biggest problems is that few of the participants are on the same page. The media's role is to sell stories; the government's role is to do what it thinks the public wants (and to minimise the costs of doing so); the police and other emergency services try to reduce the number of incidents they have to respond to; the role of the road authorities is to keep people moving with the least delay; the manufacturers and retailers want to sell more cars; the traffic safety experts and academics talk about keeping us safe. Daniel Miller poignantly states that we are in a time where industry is developing the "intelligent car" and governments are creating the "intelligent road," and so the relationship between commerce, the state, and the private driver is only going to become more problematic and make the issues of civil society, rights, and responsibilities even more acute.[53]

This suggests that we have a breakdown at the global level. When we think about health, we talk about social determinants that can be either upstream or downstream. The upstream factors are the global things, such as the economy, strategy, and policy, and the downstream factors are the more individual things, like lifestyle choices. What we do know is that a breakdown at the top, at the systemic level, results in a big disconnect between what organisations want to happen, and what actually happens at the local level.

Nevertheless, there are promising signs of a growing awareness of the need for culture change. In the United States, for example, the AAA Foundation for Traffic Safety Research hosted a national conference in 2007 to seek a way forward for improving traffic safety culture. The AAA also conducts and publishes an annual Traffic Safety Culture Index to monitor issues and changes over time.[31]

Despite the entrenched culture of cars in modern society, it is important to remember that culture is a dynamic and ever-changing process. It is not a taken-for-granted state of play that can be used to explain behaviour and enable us to just shrug our shoulders. Culture is never, in itself, an explanation. It is what needs to be explained if it is to be changed! We should consider culture as nothing more than a "tool kit" of symbols, practices, habits (norms), and views (values, beliefs, ways of life), as this can help us think about the task of change more simply. After all, culture has been changed before—seat belt wearing and smoking are two such examples. Where we need to get to in traffic safety is a place where no one will be seriously injured or killed, while behaving reasonably, just through using the road transport system. Essentially, we need to create a safety culture, with all the systems in place to support it. We are presently a long way from that point.

5 Brief History of How and Why Science Takes a Back Seat

5.1 STAGES OF OFFICIAL THINKING ABOUT TRAFFIC SAFETY

We should not forget that personal mobility through motorised transport has a relatively short history, albeit a history of rapid evolution. The horse and carriage and the bicycle began to give way to the motorcycle and the motorcar barely 100 years ago. The initial response was an immediate focus on infrastructure provision; more and better roads were needed. As motorised traffic volumes increased, an additional focus was needed on traffic management, including legislation and regulation. For example, the world's first traffic light was only switched on in 1914 (in Cleveland, Ohio),[25] while the first traffic act in France appeared in 1921 and the first in China not until 1955.[97]

Education about road rules and safe driving quickly became the principal tool within governments' crash prevention strategies based upon the (now quaint) belief that if drivers understood what was good, safe practice, they would always apply it! It was the birth of the moral approach, which has proven remarkably, and tragically, resilient. As Jennifer Clark wrote:

> The driver was expected to exhibit the traits of a gentleman: goodness, integrity, attention to duty, righteousness, courage, rationality, self-control and virtue. [This] image … informed road safety for the first sixty odd years of our motorised history so that the primary road safety message was a repetitive call for courtesy on the road.[98]

While enforcement was a supplementary strategy, it was with a "soft glove" because the police saw traffic enforcement as secondary to the enforcement of "real crime."[99] The fledgling profession of traffic engineering had no choice but to become the interpreters of human behaviour in traffic, sadly without any training in behavioural science. Traffic engineers were undoubtedly happy to be supported by road safety councils comprising eminent and concerned citizens who led the (largely) educational efforts.

With dramatically increasing rates of motorisation, particularly post-World War II, there was a rapid increase in the absolute number of deaths and serious injuries. Governments began to realise that public education to instil safe behaviours and (soft glove) enforcement to minimise irresponsible behaviours were not sufficient by themselves. The oft-chanted mantra was "the 3Es"—engineering, education, and enforcement—but each operated in splendid isolation and none had a systematic scientific base from which to draw evidence of effective interventions. It was not until governments started to systematically collect and analyse data that a scientific approach to identifying and tackling the myriad problems began.

If we, somewhat oversimply, categorise the prevailing lines of thought we can identify overlapping eras of thinking about crash causation and prevention approaches in Western motorised countries. We should also note at the outset that the thinking of traffic safety professionals has changed at a totally different rate than that of the motoring public and their elected representatives, and that a chasm exists between the two to this very day, a chasm that helps explain why our progress has lagged so far behind the evidence base.

5.1.1   1900 to 1930

Crashes were predominantly seen by both the public and the responsible authorities as inevitable and unavoidable chance phenomena, and they were considered the price of progress. As they were thought to be unavoidable, little could be done to prevent their occurrence. While transport engineering institutions were emerging, there was no scientific base for them to draw from; they had to come to grips with the previously unknown phenomenon of rapidly growing personal motorised mobility.

5.1.2   1920 to 1950

The prevailing public view was that crashes were caused mostly by poor skills or by the reprehensible behaviour of a minority of road users (a view that has doggedly persisted into the twenty-first century!). Behavioural scientists began to seek to identify the "accident-prone" in the hope that diagnostic tests applied before licencing might prove successful in keeping such people off the roads altogether or be the basis for remedial training before subsequent licencing.[100,101] While the research found significant correlations between various personality types and crash history and between prior traffic violations and crash history, the predictive ability of the potential tools was extremely poor. Excluding people from driving on the basis of test results would have had myriad false negatives; large numbers who would not have subsequently had crashes would have been excluded from the very real benefits of personal motorised transport.

In short, it was an era in which individuals were considered to be to blame for their own crashes. As a by-product, the blameworthiness explanation of crashes proved an excellent justification for road and traffic institutions and for automobile designers to avoid scrutiny over the level of safety in their designs. It is not suggested that they embraced the "blame the victim" conception consciously, but it certainly was the most cost-effective position for them at a time when infrastructure demands were high and funds scarce.

As we shall see later, while blame the victim has disappeared from the thinking of the traffic safety professionals, it is only slowly diminishing in the minds of senior managers in the key road transport agencies, and it remains alive and well in the public mind and, as a consequence, in the minds of most political leaders.

Slowly, as different types of crashes began to be examined in greater detail, particular causal factors appeared with high frequency, and these "risk factors" were considered to each require a specific preventive remedy. This was a refinement of the general blameworthiness view, with the focus now zeroing in on prevalent classes of road user fault.

5.1.3   1950 TO 1990

Gradually the evidence accumulated to indicate that crashes have multiple, interactive causes, and that a systems approach was likely to prove a more fruitful approach to prevention. The notion of a causal chain spanning an extended timeline, with injurious events being the result of the coincidence of causal circumstances—the Swiss cheese conceptual model—became an accepted conceptual approach in industrial safety long before its widespread acceptance in the field of traffic safety.[102]

Ralph Nader, who was way ahead of his time, in his seminal 1950s book *Unsafe at Any Speed*, pressed the automobile industry to accept that it should guide its design of vehicles by assuming that "fools and drunkards" will drive them.[103] For his trouble he was hounded by the industry, which only served to give enormous public attention to his book and accelerate the introduction of safety design standards on a reluctant industry!

The pioneering work of the public health specialist William Haddon Jr. brought the epidemiological discipline of host, agent, and environment and the time frames of before, during, and after the crash to the search for preventive measures in traffic safety, and the Haddon matrix revolutionised the way traffic safety professionals thought about both causes and interventions.[104] An example is depicted in Table 5.1 of ways in which measures to help prevent—or minimise injury in—single-vehicle run-off-road crashes into fixed objects. Only one of the several potential measures in each box of the matrix is shown. Once all potential measures were conceived in a

TABLE 5.1
Conceptualisation of a Haddon Matrix

	Human	Vehicle	Environment
Precrash phase	Provide off-road training in skid recovery to all drivers	Fitment of electronic stability control	Installation of audible (raised) edge lines on sealed shoulders
Crash phase	Mandatory seat belt wearing to ensure some attenuation of kinetic energy	Fitment of both front and (curtain) side airbags	Installation of wire rope barrier to substitute for collision with rigid object
Postcrash phase	Provide training in first aid for all drivers who may become first responders	Use of airbag deployment sensor and GPS to automatically notify emergency services of event and location	Seal shoulders with adequate width to enable emergency vehicle access through blocked traffic

matrix, evidence of likely effectiveness, cost of implementation, and practicality of implementation of each could be assessed to make the final selection(s).

Despite excellent advances in traffic safety science, the major focus of public policy remained on behaviour change and the search for "perfect" behaviour, especially around the prevalent risk factors of drunk driving, speeding, and aggression. This is not to denigrate the advances made in controlling unacceptable high-risk behaviours. For example, the proportion of fatal crashes that involved alcohol-affected drivers dropped by almost half in the latter half of the 1980s in two Australian states (New South Wales and Victoria) as the deterrent effect of intense random roadside breath testing had its impact.[105]

The history of attempts to reduce the high incidence of alcohol-affected driving is a poignant case study of the way legislative and enforcement experience found its way into improvements in both law and practice (see Table 5.2).

5.1.4　1980 TO 2000

Among traffic safety professionals the individual road user came to be regarded less as the predominant cause and more as the system's weakest link, with the focus shifting to a consideration of how road and traffic system design might seek to accommodate common human errors and also upon the interactions between the key Safe System components. It must be stressed, though, that this change was strongest among traffic safety professionals, and that they met continued resistance within

TABLE 5.2

History of Drunk Driving Law and Enforcement Practices

Year	Countermeasure
1909	First "driving under the influence" law passed in Australia. Enforcement was at a low level, partly because the size of the problem was poorly understood and partly because the police had to prove impairment to a court.
1955	Blood test results became admissible as part evidence of impairment; however, blood could only be taken with the driver's consent.
1961	Breath test results became admissible as well. But again, driver consent was required.
1962	The consent clauses were removed and police could require either a blood or breath test; however, they still had to prove impairment by other means.
1966	A per se law was introduced—meaning it became an offence to exceed a prescribed level of blood alcohol—thereby breaking the nexus between drinking and proof of impairment.
1976	Random breath testing was permitted by law. For the first time police could require a breath sample from any driver at any time "without cause."
1983	Research had shown that the random breath test law was not effective at the levels at which it was initially enforced. To achieve general deterrence, highly visible and highly intense enforcement was required. Once enforcement reached such levels the drop in alcohol-related fatal crashes was rapid and sustained.

Source: Birrell, J., *Twenty Years as a Police Surgeon*, Brolga Publishing, Melbourne, Australia, 2004.

substantial parts of their government agencies, partly because the prevailing blame-worthiness view remained so strong at public and political levels and partly because of the cost and potential liability implications of radically different system designs.

For the first time, nations began to formalise strategies for the reduction of disabling injury and death, particularly during the 1990s. Few of these early strategies set quantitative targets for the reduction of disabling injury and death, and the interventions proposed were those deemed politically acceptable rather than anchored in evidence of likely effectiveness. Moreover, there were no specific lines of accountability to ensure implementation.[107]

5.1.5 2000 TO PRESENT

The Safe Systems approach is being progressively refined, but the major advance in thinking has been in the mindset of those responsible for advancing safety "from homilies appealing to man's better nature through hopelessness and despair to bold political and social initiatives to a vision where no one will be killed or seriously injured."[108] It would be folly indeed to assume that widespread adoption of such a vision has been accomplished as yet. Many officials and politicians eschew accepting an ambitious visionary goal—perhaps fearing that they may be ridiculed or, of more concern, held accountable for adopting what, in their eyes, is an impossible objective. Such officials are part of the problem, not the solution.

While many governments purport to develop strategies based upon Safe Systems thinking, the action plans that follow from those strategies still very much depend upon what politicians perceive to be publicly acceptable. They remain, for the most part, demonstrably reactive.

Reinforcing the point, a leading world expert in traffic safety—Fred Wegman from the Netherlands—spent time during 2011 in South Australia examining that state's strategies. He concluded, in his 2012 report: "While the Safe System *concept* has been present in Australia for many years, its *implementation* still proves a challenge to everyone involved in road safety"[109] (emphases added).

Examples can be drawn from all Australian states to illustrate Fred Wegman's point, but we cite just two from Victoria, a decade apart, with each from a different side of the political spectrum. In 2001, the then premier was asked in a radio interview to explain the state's traffic strategy. He said, "Well, the strategy is twofold. One is to ensure that we have greater penalties; and secondly, to increase our education in a community-wide effort to say that it is the whole community's problem."[110] In 2011 the (now different) Victorian government responded to public concern over a reported increase in fatalities on rural roads by announcing a 3-year rural crash reduction strategy based upon a massive increase in the enforcement of drunk driving and speeding offences in rural areas, despite enforcement being least effective on such roads because of the vast lengths of road to be covered. (See media release in Box 5.1.) There was no mention of the options for safer infrastructure interventions on higher crash risk lengths of rural highway, despite the fact that only a few weeks earlier, the same government had publicly announced the adoption of a *new Safe Systems strategy* for traffic safety. Far from being unique, this official media statement typifies government reactions to apparent increases in road deaths that can be

BOX 5.1　PRESS RELEASE FOR ROAD TRAUMA REDUCTION STRATEGY

PRESS RELEASE FROM THE DEPUTY PREMIER OF VICTORIA, JULY 7, 2011*

Police will use tougher and more targeted enforcement methods as part of a new strategy to cut road trauma in regional and rural Victoria.

Acting Premier and Minister for Police (name deleted) launched the *three-year Regional Victoria Road Trauma Reduction Strategy* with Deputy Commissioner of Victoria Police (name deleted) today.

Mr (X) said the strategy would seek to address the over-representation of country Victorians in the annual road toll figures. "Last year 163 people tragically died in road accidents across regional Victoria, an increase of 13 per cent on the previous year," Mr (X) said.

"We need to put a stop to the carnage on regional and rural roads and *encourage drivers to take basic precautions at all times to protect themselves and other road users.*

"Research indicates that *country drivers often have a more relaxed attitude in relation to seatbelts, drink and drug driving and obeying the speed limit, which is a significant part of the problem and something that must change.*"

Mr (X) said police enforcement in country areas would vary, including increased undercover drunk-driving tests, expanded road-side drug testing and increased use of speed detection devices.

"Police will target over-represented groups such as heavy vehicles and motorcycles and clamp down on people driving under the influence of drugs or alcohol," Mr (X) said.

"There will be more breath testing and speed enforcement by undercover vehicles on secondary roads, so any driver who thinks they know the local police 'hot spots' should think again."

Mr (X) said the *strategy would focus on changing attitudes and raising awareness of the damaging effects of speeding, drunk-driving and hoon behaviour.*

"Road trauma tears apart the lives of so many families and communities across regional Victoria and it needs to stop," Mr (X) said.

*Modified only to remove personal identifiers and add emphasis.

found throughout the Western motorised world. Politicians respond to what they think the public want, particularly where it helps avoid additional expenditure from scarce funding sources.

A second, this time anecdotal, example was observed at a meeting attended by one of the authors in 2009. A local government councillor in southeast South Australia, responding to arguments by the state road safety authorities that lowering speed limits on rural two-lane, two-way roads from 110 km/h to 100 km/h would save lives, successfully resisted the proposed reductions stating that "the road toll is a fact of life." In other words, expounding the view that there is nothing we can do about road trauma, so we should not seek to change the current operation of the network!

What these examples imply is that what the science—and its application—has to say is not being heard by those who most need to listen. All three are examples of what Goh et al. called shifting the burden.[111] Instead of accepting the need for changes to the risk management system in place, the search remains for a symptomatic solution that is in line with the prevailing belief that the safety problem lies squarely in the hands of the road users. McKenna pointed out that the popularity of educational interventions lies in their low cost (relative to infrastructure investment), their avoidance of yet more regulatory control, and their plausibility given the prevalence of a blame the victim view of crash causation.[112]

What this introductory history implies is that there remains a chasm between the findings from the science and its burgeoning evidence base about the true nature of the myriad problems and effective interventions and what the public and through them, their political leaders are prepared to accept. Such a conclusion is hardly surprising, given what we have seen in previous chapters of the role of the car in modern Western society and of the level of public ignorance about the size of the problem of road trauma.

5.2   BASIC APPROACHES TO INJURY PREVENTION

The theory of risk management as applied to injury prevention in any area of human endeavour advocates the adoption of a hierarchy of controls (Figure 5.1).[113] The three most effective strategies within the six-level hierarchy are *elimination*, *substitution*, and *engineering* controls. Applied to traffic safety, an example of *risk elimination* is the separation of opposing carriageways to remove the possibility of head-on crashes. An example of the *substitution of a greater risk with a lesser one* is the installation of roadside barrier that replaces a collision with a tree (in the event of

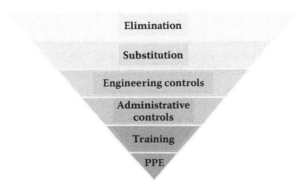

FIGURE 5.1   Hierarchy of hazard controls for the management of risk. The hierarchy of controls model depicts strategies to reduce risk, with the most effective at the top and the least effective at the bottom. PPE refers to *personal protective equipment.*

a run-off-road incident) with a collision with a barrier designed to absorb kinetic energy. An example of *reducing risk through engineering controls* is the introduction of collapsible steering columns to absorb kinetic energy when a crash occurs.

The remaining three strategies are *administrative controls, training,* and the *use of personal protective equipment.* Administrative controls include all our rules and regulations; for example, speed limits are intended to manage risk (and they do so where they are appropriate and where they are obeyed). Training is critical, as people have to learn not just vehicle handling skills, but also hazard perception skills. Personal protective equipment use is also critical and includes seat belt wearing and helmet wearing by cyclists and motorcyclists. However, we can quickly see that these three strategies rely upon the skill and motivation to comply of the people in the system, and so their success is that much harder to guarantee.

It is a fascinating fact within the history of traffic safety interventions that we have continued to rely for so long on the three least effective strategies at the bottom of the hierarchy of injury prevention controls.

The success of compulsory seat belt wearing legislation (an administrative control) is a case in point. With first-generation seat belts people had to learn how to adjust their belts (training/education); unless the belts were fastened appropriately with the buckle rearward of the hip, protection was not just suboptimal, but a new class of injury could—and did—result. Evidence of the appearance of this new type of injury led to a redesign of the belt (engineering control), and the inertia reel and the "fixed stalk" combined to improve the design dramatically, but effectiveness still required compliance with the law. While the design solution was being sought, public education was successfully used as a holding measure.[114] Before the legislation was politically possible, education campaigns served to raise voluntary wearing rates to a level that enabled political examination.

The history of traditional traffic injury prevention efforts can be summed up in the following three approaches, the 3Es of education, enforcement, and engineering[32]:

1. **Educate:** Persuade persons at risk to change their behaviour. The educational approach has developed considerably beyond its early general exhortations to drive carefully to targeted campaigns to change specific risky behaviours—for example, to encourage seat belt and helmet wearing and to make drunk driving socially unacceptable.[115] Education and training have enjoyed widespread popularity as safety measures, and considerable success in specific cases, but they are clearly not a panacea. Public education has proved its greatest value as a way to raise awareness about a pressing safety issue, as a contributor toward building a political consensus for change, and as an essential ingredient of support for new legislation or intense enforcement.

2. **Regulate and enforce:** Mandate behaviour change (for example, the law that made it illegal to drive when drunk, and was gradually refined to change the offence to one of exceeding a prescribed concentration of alcohol in the bloodstream). Regulatory policy options have historically been unpopular and strongly resisted. In Australia we seem to have forgotten

how hard it was to mandate seat belt wearing and to obtain per se laws for drunk driving, but accounts of the history demonstrate how controversial they were at the time they were being considered.[106] The popular media often ridicules the regulatory option with cries of "nanny state" and appeals to the acceptance of more personal responsibility to solve social problems. Broad acceptance of change can, in some cases, take two or more decades. The gradual evolution of regulations around smoking in public spaces bears witness to the time social change can take.

3. **Design:** The provision of lower-risk situations through vehicle, road, and traffic system design came slowly to road transport, far more slowly than it did in industry. The designers of a factory were forced to accept responsibility for a safe environment and were prosecuted for obvious breaches, such as failing to place a guard over the moving parts of a machine. In road transport, the system designers and operators have, until quite recently, avoided public scrutiny for Safe System design, frequently on the grounds that extant system design is as safe as can be afforded, and that human error is almost always found in the causal chain when incidents occur.

6 Evolution of Safe System Thinking

6.1 INHERENT UNSAFETY OF OUR PRESENT ROAD USE SYSTEM

Very few of us stop to consider the basic characteristics of our road transport system. When we visit low-income countries we note with a wry smile that motorcycle helmet and seat belt use are rare, that walking on the road is common, and that when sidewalks do exist, they tend be filled with traders and their goods. These are things we, in motorised societies, no longer tolerate, but we do tolerate, unquestioningly, seriously unsafe designs, even though they are within our understanding and in front of our eyes every day. We tolerate and blithely accept them because we have grown up with them and we believe that we are totally in control of our own destiny on the roads, and that serious crashes only happen to unskilled or reckless others.

We doubt that Noel and Jan would accept this view, and thus the tragic outcome of their journey south from Queensland (Chapter 2).

Consider the four most common types of crash that result in fatalities and serious injuries in Australia (percentages are only for fatal crashes and use 2006 data):[116]

- Single-vehicle, run-off-road crashes (37%)
- Crashes between vehicles at intersections (14%)
- Head-on crashes between two vehicles travelling in opposing directions (14%)
- Vehicle-pedestrian crashes (16%)

Even a preliminary consideration of the nature of these crash types will indicate how unsafe much of the road network is. We accept, for example, the risk involved in a road design that permits two vehicles travelling in opposing directions to approach each other at the speed limit with scant horizontal separation. Would the driver of a vehicle travelling on a two-lane, two-way rural road at a speed just within the legal 100 km/h speed limit be as comfortable sitting on a chair just beside the white centre line marking and observing vehicles passing him in the adjacent lane at a speed of some 100 km/h only centimetres away? Of course, as a driver he has the protection afforded by the crashworthiness of his vehicle, but if two vehicles approach at legal speed, the impact forces, should a crash occur, are almost certain to exceed those of the design level of protection. More than two decades ago, Ezra Hauer chided the road and traffic engineering profession for their "communal learning disability," which blinded the profession and made questionable standards and practices "believable by uncritical repetition." He concluded that the (then) "existing level of safety was allowed to materialise in a largely unpremeditated manner."[117]

This reliance on "perfect" behaviour by drivers and the notion that a small piece of white centreline paint on a sealed surface is sufficient to prevent head-on crashes is indicative of what we accept as normal conditions. It is, of course, our experience since childhood—we have grown up with it, and that probably explains why we do not question it. The real issue is one of risk management and whether the frequency with which this daily risk results in a crash is such as to justify the cost of replacing all two-lane, two-way roads.

The Swedes mitigated the head-on crash risk by installing wire-rope barrier in the centre of stretches of such roads with excellent safety benefits and highly favourable benefit-cost ratios (Figure 6.1). Design solutions can be affordable, but are often rejected out-of-hand as impracticable and unnecessary. Where the funds for infrastructure safety improvements simply cannot be found in the short term, lowering speed limits to match the current level of protection, and rigorously enforcing them, provides a viable option.

A similar situation can be observed with roadside objects where a simple error leading to a vehicle leaving the carriageway out of control will often result in a side-impact crash into a tree or pole (Figure 6.2). Such a crash at impact speeds of more than 40 km/h is often fatal, and yet these conditions exist throughout much of the Australian rural highway network.

With intersections, it is known that collisions where the impacting vehicle is travelling at greater than 50 km/h and collides into the side of another vehicle are likely to lead to fatal outcomes. Again, there are many intersections where speed limits on the intersecting roads are 60, 70, or even 80 km/h in Australian cities (the latter in particular resulting in travel speeds through urban signalised intersections unmatched in most parts of the motorised world), and on country roads usually 100 or 110 km/h. These are situations where serious casualty outcomes are almost inevitable whenever a crash occurs.

FIGURE 6.1   A centre-road installation in Sweden. Such installations have virtually eliminated fatalities in head-on crashes on treated roads.

FIGURE 6.2 Side impact with a tree. (Courtesy of Monash University Accident Research Centre.)

With pedestrians, the science tells us that vehicle travel speeds of 40 km/h or preferably 30 km/h will generally not result in fatal outcomes if a pedestrian is struck. And yet it is still quite common to see pedestrian areas in shopping centres or other busy precincts where speed limits are 60 or even 70 km/h.

All of these situations represent a failure by our society to be aware of the risks and to provide appropriate levels of safety. The word *failure* is, of course, pejorative in this context, and it is more accurate to say that our design policy deems the risks acceptable. We used the word *failure* deliberately because the debate around design options is rarely thorough, and the conclusions remain anchored in the prevailing view of blameworthiness. Note that we also said it is a failure *by our society*, not by government, because governments respond to their understanding of what society wants. We have an unsafe road and road transport system because we have grown up with its evolution and accepted its consequences, as vehicle power and top speed capability have increased inexorably and we have not demanded a system that is commensurately safer.

6.2 EVERYDAY ERROR VERSUS BLAMEWORTHY BEHAVIOUR

This lack of understanding of the inherently unsafe nature of much of the network (even when travel speeds are below prevailing speed limits) is reinforced by the regular emphasis in the media on sensational crashes where aberrant behaviour is evident. The "blame the road user" approach and the implied ongoing need for drivers to perform perfectly has been a stock standard way of reporting on road safety by the media (as we saw in Chapter 4).

Necessary and understandable police comment at the scene in the media about crashes implies to viewers that improper behaviour is always the usual cause of crashes. These immediate responses by people trained to seek fault further reinforce the popular view.

Police reports of crashes almost inevitably identify a specific human behaviour as the final link in the causal chain. This is hardly surprising, as the human is the only active element in the open-loop system that is road use. Moreover, human behaviour is what the police are trained to investigate. What does the scientific research tell us about the involvement of illegal behaviours (breaking road laws/road rules) in serious casualty crashes?

Scandinavian research based upon an extensive literature review suggested that in about half of fatal crashes and over two-thirds of nonfatal injury crashes driver error that was *not* related to breaking the rules could be identified. In short, error is common and typically legal![14]

Comprehensive studies have been undertaken in South Australia to further test this finding. The Centre for Automotive Research (CASR) at the University of Adelaide examined all the information contained in the state coroner's investigation files for fatal crashes in 2008.[118] As the coroner examines all fatal crashes, this was a 100% sample for that year. CASR also undertakes on-scene, in-depth crash investigations as part of its ongoing research program. It examined all the data collected from its sample of nonfatal casualty crashes for the same year as the fatal cases. The analysis included 83 fatal crashes, 272 nonfatal metropolitan injury crashes, and 181 nonfatal rural crashes. Very few nonfatal crashes (3% metropolitan, 9% rural) involved extreme behaviour by road users, and even in fatal crashes, the majority (57%) were the result of system failures. This means that improvements to the road transport system can be expected to be much more effective in reducing crashes than concentrating on preventing extreme behaviours.

This work provides evidence that more than 50% of driver fatalities and about 90% of driver serious injuries arising from rural road crashes did not involve any breaking of the law/road rules. Once we understand error better, we can design a safer system.

That we have been slow and imperfect in our efforts to match traffic law and traffic management to normal human behaviour can be illustrated through a case study of risk management at intersections in Australia.[119]

The original law governing vehicle priority at uncontrolled intersections was that in the event of a potential conflict, the vehicle "on the right" (Australians drive on the left) has right of way. This simple rule saved authorities the cost of installing signs and pavement markings to control intersections without high volumes of traffic. It also simplified the task of police and insurance companies since the (legally) guilty party was immediately obvious. The rule ignored, of course, the complexity of the judgements involved, and intersection collisions were rife (Figure 6.3).

One downside of this simple rule was that traffic on major roads was often interrupted by traffic on minor roads intersecting with the major road. Given the objective of maximising traffic flow, the give-way-to-the-right law was modified for high-volume roads by signing them as priority roads. This ensured that vehicles on

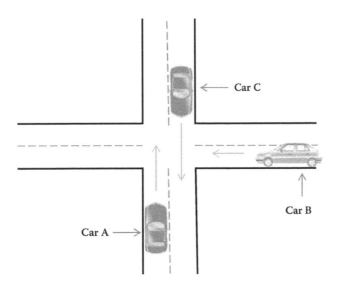

FIGURE 6.3 Decision complexity at intersections. A driver (A) approaching an intersection had to decide if any vehicles approaching the same intersection from his right (B) were in potential conflict, requiring notoriously unreliable judgements of distance and relative closing speeds. Decision complexity was further compounded if a third vehicle (C) was approaching on the same road as (A), but from the opposite direction, to which (B) may have to yield. Would (C) get there in time to "save" (A) from having to yield? The rule encouraged aggressive driving because, where uncertainty exists, you can speed up to "beat" the vehicle to which you may otherwise have to yield. (Adapted with permission from Johnston, I., Highway Safety, in *The Handbook of Highway Engineering*, ed. T. Fwa, CRC Press/Taylor & Francis, New York, 2006.)

minor roads intersecting with major roads had no priority over major road vehicles approaching from their right. The signs were colloquially (and very appropriately) known as rocket road signs. Collisions did not decrease; indeed, it appears that risk taking increased as speeds on the major roads rose and drivers on the minor roads became frustrated with the time taken to enter or clear the major road and took greater risks.

Finally, the give-way-to-right-rule was scrapped and all unsignalised intersections were treated with either STOP or GIVE WAY signs and with pavement markings. The cost savings gained by avoiding formal controls could no longer be justified! This case study beautifully illustrates the influence regulatory and traffic management decisions have on fostering safe or unsafe behaviour. It is not that drivers flout laws; they take risks as they understand them.

One can (simplistically) divide road users into four groups, arranged in a pyramid (Figure 6.4). At the base are the majority who comply with laws and informal norms almost all the time. Next are those who also want to comply but occasionally transgress, usually without serious intent. Then there are those who fairly regularly and quite deliberately break the rules but can be deterred if probability of detection is sufficiently raised. Finally, the tip of the iceberg comprises the small minority who "don't give a ****" and have no desire to comply.

FIGURE 6.4   Pyramid of road user groups.

Shinar, in the United States, wrote that social maladjustment is reflected in driving maladjustment.[13] Broughton took a sample of records for over 50,000 drivers in the United Kingdom and found a "strong and dramatic" relationship between traffic violations and nontraffic offences such as theft, drugs, and violence. Those with more than 4 traffic offences had 21 times as many nontraffic offences as those with no traffic offences.[120] It should be clear that a focus on the tip of the behavioural pyramid cannot be sufficient. We have a problem with antisocial driving behaviour precisely because we have the full spectrum of society operating as road users, and we have an overarching problem with the antisocial behaviour of a small minority of the population.

We reiterate, however, that it is crucial to understand that antisocial behaviours by a minority of road users account for a relatively small proportion of the aggregate of road trauma. The evidence presented earlier in this section is clear but not widely known.

Drawing from the work in industrial safety of James Reason,[121] we can distinguish between the following:

- **Errors:** Unintended failures of planned actions, for example, a rear-end crash occurring because the driver spent too much time scanning the cross-road for possibly conflicting traffic at an intersection.
- **Lapses:** Attention or memory failures, for example, a driver selecting the wrong turn lane at an intersection.
- **Violations:** Deliberate deviations from the rules.

A lot of each type of behaviour is system related. The violation of red light running, for example, is closely related to perceived waiting time—red light running at intersections has been observed empirically to be greater in peak periods than in nonpeak periods, and on workdays compared with weekends, and during periods of congestion.[13] Transient risky behaviour is typically rewarded by the immediate perceived gain.

Driving is essentially an unstructured task: "a complex human activity requiring adaptive behaviour."[122] The road rules leave lots of latitude, the available information is often informal, and skill acquisition is primarily learning by doing. The latter illustrates why the mandating of a quantum of supervised experience before going solo, as required in many of the graduated licencing schemes introduced around the world, is such a powerful measure.

Error is a natural part of any operation involving humans. Error is adaptive in the learning process. It is astonishingly frequent in everyday driving, with only the tiniest minority of errors resulting in a crash.[74,123] A study of crashes will tell us a certain amount; a study of near misses will tell us more because a near miss is an event that almost had adverse consequences. What might a study of driving error tell us? Research is underway to tap this gold mine of information that will benefit system designers in their search for error-reducing designs.

Amalberti cites the Aviation Safety Reporting System as an excellent example of a no-blame reporting system that generates over 60,000 reports per annum of mistakes made by air and ground crew, the vast majority of which did not result in incidents or near misses.[124] If we are to study error, we need to study it in its context—for example, over 40% of all serious casualties occur at intersections.[31,125] Let us study error at intersections with a view to providing controls to ensure errors don't lead to crashes or providing aids to decrease the likelihood of the safety-critical errors. There can be no doubt that solutions exist.

People clearly make many ordinary mistakes when driving, and this can lead to serious casualty outcomes, far more frequently than those arising from illegal behaviours. Of course, we must continue resolutely our determined efforts to reduce aberrant illegal and unsafe behaviours; society must be protected from its miscreants. But it must not be our sole focus; we must put at least equal emphasis on a redesign of our systems to decrease the likelihood of the most common everyday errors that occur. Technology will certainly assist this process. However, we cannot expect human beings to behave perfectly at all times even when they are behaving legally.

These shifts require and depend upon a major increase in public and practitioner understanding if we are to progress this emerging conceptual approach.

Thanks to the Swedes and the Dutch, we have arrived at Safe System thinking. The simplicity of early heuristics such as the 3Es, *education*, *engineering*, and *enforcement*, has been discarded as we have come to understand that we need truly complementary—not competitive or narrowly focussed or single professional function-based—intervention strategies. Effective intervention implies proactive innovation.

6.3   DECISION-MAKING CONTEXT FOR PUBLIC POLICY DEVELOPMENT

On the whole, politicians respond to their understanding of what the people who elected them want or will accept as policy on social issues, coloured very strongly by the perceived cost of the options before them. This understanding is mediated by "gatekeepers"—senior government bureaucrats, advisers, and lobbyists who purport to interpret the available evidence. It is not uncommon for "solutions" that are potentially controversial to not even be brought forward for political consideration. This may reflect, on the part of the gatekeepers, a lack of confidence, a lack of commitment to a cause, a desire to avoid accountability, or a wish for a more comfortable existence. It is so much easier to stick with the conventional wisdom. In the brilliant British television series *Yes Minister*, Sir Humphrey Appleby, the supreme controlling bureaucrat, was wont to say, "That would be a courageous decision minister!" whenever he wished to dissuade the minister from a line of action.[126]

Potential medium-term gains (beyond a 2- to 5-year time horizon) are usually heavily discounted by political leaders and many senior bureaucrats who may be on short-term (often 5-year) employment contracts. There is limited perceived payoff for them in personal terms to recommend radical solutions to a traffic safety problem that lacks political saliency. In our experience, often the response to discussion about a policy option that is progressive will be "I don't think my minister could support that," which on further probing (when possible) often leads to an admission that the officer has personal qualms about the suggestion, with his or her minister not even being let into the conversation on the issue.

This behaviour is, it is asserted, widespread in Australia and New Zealand and in many motorising countries in their region; it is tied up with values and competence and a risk-averse philosophy and has a major effect upon public policy development and outcomes, especially in the challenging area of traffic safety. When the gatekeepers are not as expert in the subject matter as their senior advising staff, then suboptimal conveyance of any advice to ministers is much more likely.

Robust policy adoption by government relies upon robust gatekeepers who provide strong, well-developed primary issue advice, taking care to provide associated advice about managing unhelpful impacts of the policy, thereby increasing its chances of adoption by the political level. Progress of the transformational change required to eliminate disabling injury and death from road transport requires committed professionals to push for public policy change and associated funding based on evidence in as vigorous a manner as possible. So often the gatekeepers apply an "excess of realism," which prevents many of the potential projects proceeding. That is, the "illusion of hope" that drives advocacy for change is dashed on an ongoing basis.

This is usually all too easy for the realists. Rarely, for example, do treasury officials have to stand in front of TV cameras and justify to the public why they did not support an evidence-based proposal for an investment in a policy or project. The public is never aware of those thumbs-down decisions made out of sight with considerable frequency in any government. Yet they cost lives. Achieving successful change—which of course relies on a healthy mix of both approaches—does rely on aspiration. This tends to be overlooked in risk-averse public sector operation with the increasing focus on populism by politicians in recent years making difficult but evidence-based policy adoption less common.

As superbly expressed by Ezra Hauer, "A broad class of professionals, those who influence the future of road safety, needs to be trained.... Above all, they need to be freed from the constraints imposed on them.... The best interest of society is to move toward the gradual establishment of the rational style of road safety management, it is the engineer's professional obligation to promote this societal interest."[127]

6.4 MAKING TRADE-OFF DECISIONS

Before turning to a description of the Safe System, it is worth exploring further the typical policy context in which decisions about road transport system design and operation are made. We concluded Chapter 3 with an outline of the myriad objectives confounding the development of transport policy. Let us now develop this a little further.

The triple bottom line we often hear about in transport means a system that maximises economic return, maximises safety, and minimises adverse social and environmental impacts. These objectives are typically seen as competing, requiring trade-off decision making. Trade-off decisions are typically made on the basis of cost-benefit analysis—dollar values are assigned, for example, to travel time, to pollutants, and to injury (but rarely to social justice outcomes), and then the costs of a proposal are weighed against the benefits, with only proposals showing positive returns getting approved. Occasionally, necessity is the mother of trade-off decision making—witness the primacy of emission controls on vehicles in California as a response to appalling air quality where California moved way ahead of U.S. national standards. This is important evidence that if a specific issue is sufficiently salient, it will be dominant in the decision-making process. We argue a case that disabling injury and death is such an issue.

There are vigorous academic debates around the assignment of dollar values to costs and benefits on several levels. First, we include in our equations only those things that we can measure or estimate, yet there are many unknowns in assessing environmental and social impacts in particular. We cannot be confident that we have accurately identified all the key elements or that by simply excluding those that we cannot measure, we have not seriously biased the outcome. As Dunn pointed out, we risk solving the wrong problem by applying incongruent methods.[128]

Second, there are debates about how best to establish the true costs for individual elements—for example, on injury, whether we should count all the costs actually incurred for injury at various levels (loss of earnings, hospital costs, damage repair costs, etc.) or whether we should estimate the community's willingness to pay to save a disabling injury or a life.

In New Zealand, the government not only required positive benefit-cost ratios before a project could proceed, but also mandated that the economic returns must reach a prescribed positive level. When this policy was introduced, safety projects became uncompetitive. The traffic safety specialists made a successful case for a switch from the direct cost method to the willingness-to-pay method for estimating the dollar value of a statistical life. The switch resulted in an approximate doubling of the value of a life and got a lot more safety projects "over the line" in the benefit-cost competition. Sadly, traffic safety advocates in New Zealand currently bemoan a failure to adequately adjust overtime, the values derived of this paradigm shift of well over a decade ago.[129]

There can be no doubt that whatever the decision-making method, games will be played by advocates seeking decisions in their favour. This is not to suggest that the New Zealand safety specialists were behaving inappropriately; they were simply putting their best foot forward, just as the advocates of congestion amelioration proposals do.

Similarly, there are debates about the value of time—is recreational travel time the same as commuting travel time or the same as business travel time? While these debates are interesting, they miss the central question: Can public decision making be reduced to a simple dollar value that results from attempts to turn all the products and consequences of road transport into quantitative monetary units and combine them in a single equation? As with the continuing debate over our common usage of the term *accident* when we should have used *crash*, we spend a lot of time on unproductive issues. The real question here is an ethical one—the sanctity of life. There is a major place for the monetary assessment of benefits and costs when choosing between policy options, but it must not be the sole criterion.

Rose, with his tongue firmly in his cheek, took the economic rationalist argument to its cynical extreme by assessing the benefits (to the economy) of saving a human life at various stages of the life cycle. He assigned no value to an infant, as an infant costs little to produce and little to replace, and a low value to people over 60 because the loss of a relatively small number of productive years just about offsets the cost of aged welfare and care. Young adults are, on the other hand, a grievous economic loss, so we should focus our (cost-beneficial) intervention efforts only at this level![130]

What philosophical assumptions must be accepted if you stick with the conventional benefit-cost ratio? The key one is that life and disabling injury are reasonable prices to pay for progress. The Swedes, as we have already seen, do not permit safety to enter the benefit-cost analysis. No proposal that increases unsafety will be entertained. Let us examine what they have done and, in so doing, acknowledge our Swedish colleague Professor Claes Tingvaal, one of the pioneers in the Swedish progress on traffic safety thinking.[*]

> The balance between safety, environment and accessibility is of course a never ending story. Sweden has in the last two years found a new way to express transport policy. In summary it is: "accessibility can only be developed within the framework of safety and environment". The absolute position expressed in this statement is that, in the long run, safety is paramount.
>
> Mobility and accessibility make up the functionality of the transport system, but there is a parameter, safety, like many other elements in a society, that is not a variable in an equation, but has threshold limits that cannot be exceeded.
>
> This is all fine, but reality is another thing and of course this shift is a gradual process. The new speed limit system is a good, although slow, way which demonstrates how mindsets shift. In the system, 80 km/h is the maximum speed for an undivided road (unless there is very little traffic volume), and this will be absolute—even if calculations would show that 90 or 100 km/h would be beneficial.
>
> Calculations such as this are not the toughest enemy of safety, rather it is the politics. Sweden is gradually changing its approach to permissible travel speeds and therefore to the setting of speed limits on new roads, or determining the investments necessary to modify existing roads to allow higher limits, based on the new speed limit system.

---

[*] We are indebted to Professor Claes Tingvaal (director of Traffic Safety at the Swedish National Road Administration) for describing the current Swedish position and allowing us to include his description.

This is where the safety benefits will be realised as the mobility needs will then be the deciding parameter for the investment necessary to have safe higher travel speeds. Is it worthwhile to invest in safety solutions to increase mobility? This is the right question, but it has taken many years to have this rather natural logic understood and accepted. The transport organisation in Sweden now has responsibility for rail and road transport safety. This logic of investment in safety solutions to increase mobility is long established for rail safety. It has been agreed that in an integrated transport system, rail and road should have the same safety paradigm, and that is a good thing.

6.5   SAFE SYSTEM APPROACH

The Safe System approach is a fundamental shift from traditional traffic safety thinking. It reframes the ways in which traffic safety is viewed and managed. Its aim is to support development of a transport system better able to accommodate inevitable human error. The recognition that humans do make, and will continue to make, errors of judgement as road users is one of the core shifts in thinking.

Adapting our road transport system to respond to inevitable human error can best be achieved through better management of crash energy, so that when an error that leads to a crash occurs, no individual road user is exposed to a level of crash forces that exceeds the capacity of the human body to withstand.

The Safe System relies on considering a number of key cornerstones, which contribute together in any crash to the severity of the outcome. However, the benefit of the Safe System approach is that action taken in each cornerstone area can often be applied together in a complementary way. Careful planning of the potential individual adjustments that could be effectively applied at particular locations, or preferably at locations or along lengths across the network, can maximise the traffic safety performance benefits achieved.

It is a key challenge for every traffic safety agency to establish a deep understanding of the critical factors in the road and traffic environment, vehicles and travel speeds that lead to the four most prevalent serious injury crash types. Then the challenge over time is to implement innovative and appropriate measures to take advantage of the opportunities to reduce kinetic energy exchange when crashes do occur.

The Safe System cornerstones that are conventionally discussed are

- Safer vehicles (including system access requirements)
- Safer roads
- Safer speeds
- Safer road users (including licencing or system access requirements)

These cornerstones should be supported by

- A higher standard of postcrash care
- Improved traffic safety management, the underpinning enabler of delivery of successful traffic safety performance

While ongoing efforts are certainly required to achieve alert and compliant road users, major continuing benefits will be achieved through focussing on road network safety improvements (achieving forgiving infrastructure) in conjunction with reviews of posted speed limits (to be set in response to the level of protection offered by the road infrastructure) and by the progressive introduction into the fleet of modern vehicle safety features. Enhanced vehicle safety features are becoming available at a remarkable rate. Encouragingly, at least in Australia, the public is, underpinned by extensive public education, demonstrating a growing market for such features.[131]

Addressing the opportunities available through interactions between these system elements (roads, vehicles, travel speeds) to reduce combined crash energy will provide major benefits. As a consequence, the road transport system will be made fundamentally safer.

The Safe System approach recognises that the responsibility for safe operation of the network is shared between many individuals and organisations, including those who provide the roads, set the speed limits, make the laws, provide the vehicles, make the land use planning decisions affecting traffic flows and roadside access, use the network, enter contracts for supply of transport services, enforce compliance, employ drivers to use the road network in their work, operate the emergency health system, and more. It is important to stress that the individual road user remains a critical part of shared responsibility. Each user has a responsibility to obey all traffic laws at all times and to ensure fitness to drive by avoiding impairing substances and managing fatigue. However, as the University of Adelaide research shows, individual responsibility is important but not sufficient to achieve a truly safe road transport system.

This changed view of road user responsibilities away from a solely blame the road user emphasis is a key feature of a Safe System approach. An important set of challenges is involved in determining and monitoring ongoing performance against the accountabilities of all players. Sadly, when the media, and all too often politicians and senior bureaucrats, talk about greater responsibility, they mean that road users must accept the major share of behaving responsibly—each must be accountable for his or her actions. Indeed, we must, but we should not have to do so in a system full of situations of flawed design that increase the probability of error.

As Belin points out, the Swedish Vision Zero "as a public policy envisages a chain of responsibility that both begins and ends with the system designers."[132] Of course, the accountability and financial implications of such a policy are exactly what make it so difficult for governments to accept.

Hugh McKay, the prominent Australian social commentator, told the 2011 Australasian Road Safety Conference in Perth: "Behaviour change requires a change of environment."[133]

Public information and education is necessary to inform the public about why rules are important and what the consequences of being detected disobeying these laws will be. However, to achieve maximum compliance, it is essential to deter drivers from breaking these laws in all circumstances. Many people will comply, but a proportion will get away with what they can get away with, and many people will usually not be focussed on readily changing their behaviour.

That is why the probability of detection, certainty of a penalty if detected, and deterrent level of penalty are so important. Many well-meaning community members have some difficulties with accepting—or indeed understanding—deterrence theory, but those who have had exposure, for example, to the immediacy of reductions in drunk driving fatalities as police substantially increase random breath testing (RBT) intensity in a region recognise only too well the sensitivity of the driving population (and the level of related deaths and serious injury crashes) to increased general deterrence as represented by increased widespread visible enforcement.[134]

This is an absolute tenet of improved traffic safety—at least into the medium term, until emerging vehicle technologies such as universal alcohol detectors/interlocks and mandatory speed limiters will be capable and widespread enough to require and achieve an almost universal level of behavioural compliance.

However, we should not be complacent about the inevitability of this technology coming into effective operation soon. The long-established technology of speed limiters for trucks and the abuse of this mandatory technology by certain rogue heavy-vehicle operators in Australia recently are salutary tales (see Box 6.1). This indicates the importance of effective deterrence to ensure that safety technologies are not sidelined. If a proportion of the population think they can get away with something that makes them more competitive in practice or in their minds, then they will do so.

BOX 6.1   SPEED LIMITERS IN THE TRUCKING INDUSTRY

EXTRACT FROM THE SYDNEY *MORNING HERALD*, AUGUST 27, 2011

Road safety regulators have warned major retailers … that they may end up before a court for placing unrealistic demands on transport companies and encouraging drivers to drive dangerously. A three day crackdown on rogue truck drivers, the fourth in six months, ended yesterday with … investigators finding more evidence that speed limit tampering is still rife in the trucking industry.*

Since February, 50 trucks with illegally modified speed limiters have been found travelling on the roads, sometimes at up to 142 km/h.

Under 'chain of responsibility' laws more than 1000 charges have been laid against the directors and managers of four trucking companies.

However … the next round of criminal charges arising from yesterday's crackdown would target the consignors and consignees who place unrealistic demands on truck drivers such as demanding they be at certain places at certain times or risk losing contracts.

[A Transport Workers Union] recent survey of 715 drivers found that 27 percent felt they had to drive too fast and 40 percent felt pressures to drive longer than legally allowed. Many said the pressure came directly or indirectly from the client. Since February … have inspected 1092 trucks and issued 500 drivers with fines and court attendance notices.

*Authors have edited the text to remove personal identifiers.

The Safe System approach should also recognise and seek to build upon opportunities for improved alignment of traffic safety policy with other societal goals—for example, important synergies exist with environmental protection and energy conservation policies, with occupational health and safety policies that target safer work-related driving, and with broader transport and travel policies that seek to improve broader economic outcomes, travel cost efficiency, and health and well-being through encouraging walking, cycling, and the use of public transport.

Approaches to involve all the system designers, to motivate them to achieve their potential in supporting improved safety of the system plus the provision of tools to guide and assist their contribution to that task, are essential means to develop the movement toward Safe System achievement. A Safe System approach requires community understanding and support. Only then is there a likelihood that the community will demand and expect Safe System improvements.

The impact of the Swedish and Dutch traffic safety visions on other countries has been profound. "While the escalated level of ambition (zero deaths and serious injuries) represents a radical shift within the road sector, these targets can be viewed as consistent with the safety expectations acceptable in other modes of transport (for example the aviation and rail sectors). What was initially seen as radical and unachievable has increasingly become the benchmark for acceptable road safety results."[135]

Safe System concepts have been adopted by other countries, including Norway, Australia, and New Zealand. They are supported by the WHO, the FIA Foundation, the iRAP organisation, the Global Road Safety Partnership, and the World Road Association (PIARC). Importantly, as the adopted basis for the UN Decade of Action, they are forming the basis for many national traffic safety strategies. Concept adoption and effective implementation are two different things, however, as we have already noted. We need to keep remembering Fred Wegman's caution that implementation remains our greatest challenge.[109]

We stress again that the adoption of a long-term goal to achieve a safe system requires a vision for the *ultimate elimination* of death and serious injury on the road system—an outcome of achieving a system that is fundamentally safe.

While it is not necessary to specify when the elimination condition will be reached, it is a strong statement about the unacceptability to a society of serious casualties. Its adoption also informs communities that this is a reasonable long-term expectation, and it responds to a strong and growing market for safety as community awareness grows, best exemplified in the Volvo statement "no Volvo occupant or pedestrian will be killed or injured in a post 2020 Volvo."

What does this approach—adopting a long-term elimination vision for road crash fatalities and serious injuries—require of decision makers in the interim period wishing to progress toward the long-term goal?

This is where a Safe System-based approach begins to provide some bite. It is no longer sufficient to keep on doing those things that will bring some incremental improvement in unlinked areas of the road traffic system. Focussing in isolation on reducing drunk driving or on treating road crash black spots is clearly insufficient. Given that we seek elimination, we need to work toward a safe road transport system, based on the underlying Safe System principles, in order to deliver on the long-term ambition. This shapes the directions we need to follow, the steps we take. We need to consider the serious casualty crash risks that exist across the whole road network and progressively introduce measures that will eliminate the serious casualty crash risk—generally in the most cost-effective order of priority, but always in ways that do not leave high-risk islands of the network stranded and untreated.

6.6 INSTITUTIONAL MANAGEMENT REALLY MATTERS

Traffic safety research has a proud reputation with many critical pieces of innovative research, including quality evaluations of experimental initiatives leading to well-known step reductions in serious casualties. What has received relatively little attention, unfortunately, is how to take proven research and implement the accepted findings in practice to gain the benefits.

In some ways, knowing what will be effective is only a small (yet important) part of the overall picture. Without the management skills, technical and political knowledge, and commitment and persistence required to implement these skills and knowledge, change does not take place (Figure 6.5). So often during a traffic safety management capacity review the client country practitioners will intone: "We know what it is we need to do, as you are pointing out. We understand that. But can you teach us *how to do it*?"

FIGURE 6.5 Safe System in road safety management. (Courtesy of Eric Howard.)

There are four major levels of accountability for effective traffic safety management:

- **Strategic:** At this level, responsibility must be accepted for adopting Safe System principles, changing the institutional climate and public understanding to facilitate the implementation of a new approach, managing the decision making and consultation arrangements, setting targets, prioritising problems, and establishing a formal countermeasure choice process.
- **Tactical:** Arranging funding, modifying institutional structures, facilitating necessary supporting regulations, public and political advocacy, and establishing appropriate performance measures.
- **Operational:** Effective implementation on the ground.
- **Refinement:** Evaluation at process, output, and outcome levels, with diagnostics to provide feedback (interventions don't either work or not work—they work to a degree under certain circumstances).

It is rare to find nations with accountabilities appropriately assigned at each level. Sweden and the Netherlands are probably the closest.

The need for help in traffic safety management capacity building is a common theme from low- and middle-income countries. Sadly, it should also be, but is not, a common theme in Western motorised nations. The comfort of analysis within the detached research institution needs to give way to allow some research into what is needed to guide management decision making within the furnace of policy making and budget setting within existing government organisations—with all the accompanying pressures of the associated accountabilities for performance.

Thomas Friedman,[136] in reviewing a recently published book, *Why Nations Fail*,[137] asserts that the major differences between countries' political and economic development can be attributed to the comparative health of their key institutions. Nations thrive when political and economic institutions become inclusive and fail when institutions concentrate power in a few hands.

The comparative health of the institutions that support traffic safety outcomes is of course known to be crucial in determining the relative level of traffic safety performance that is achieved by any country compared with its peers. Institutional health remains a major potential barrier to improved traffic safety performance for many high- and most middle- and low-income countries to varying degrees. Traffic safety has steadfastly stood alone, quite separate from developments in safety management in other transport modes or in industrial safety. A recent review of the effectiveness of safety management systems undertaken by the Australian Transport Safety Bureau comprehensively covered aviation, rail, and marine safety systems with no reference to road transport![138]

Recent work in this area has drawn upon practitioner management experience in Europe and Australasia and is reflected in the *Towards Zero* report[135] and the World Bank guidelines for conduct of capacity reviews.[139]

The *intervention set* of the road safety management system (as represented by the middle layer of the pyramid in Figure 6.6) has three main components:

- Planning, design, operation, and use of the road environment
- Entry and exit of vehicles and drivers from the system
- Recovery and rehabilitation of crash victims

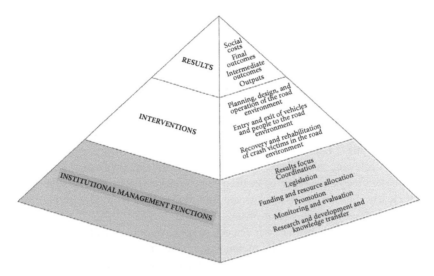

FIGURE 6.6 Institutional management in road safety. (Reproduced with permission from Bliss, T., and Breen, J., *Implementing the Recommendations of the World Report on Traffic Injury Prevention*, World Bank, Washington, DC, 2009.)

All of these areas of implementation require considerable skill to devise interventions that will deliver benefits cost-effectively, but they also rely totally on the issues—*the institutional management function* capacities—identified in the lower level of the pyramid, if they are to actually happen.

These institutional management function activities require key systems, processes, and management competencies to be in place within the government traffic safety agencies—as these impact upon capacity to deliver change (through government and nongovernment sectors) in the level of safety for the operation of the road network. They underpin traffic safety success in a jurisdiction and offer an explanation for (most of) the difference in traffic safety performance between otherwise similar jurisdictions.

The quality of the institutional management functions of government agencies in any jurisdiction can be assessed by

- The strength of relationships between government officers and the political level and the coordination between the agencies (and between their ministers)
- The clarity of agency roles and responsibilities
- The accountabilities for delivering agreed strategy and targets
- The effectiveness of deterrence through current legislation and the quality of systems for offence recording, crash data, and enforcement
- The level of supportive funding and resource allocation
- The quality of "managing up" activity—the commitment to advocacy by middle management to senior officers, and in turn by these senior officers to politicians
- The quality of knowledge as expressed in technical standards (for infrastructure, enforcement, vehicles, and licencing)

- The transfer of that knowledge and the research and development settings
- The monitoring and evaluation of interventions and outcomes

Accountability is crucial at all stages, but never more so than at the top of the key institutions. For example, the contract of employment of a chief executive of the lead traffic safety agency in one Australian state included, as the first of its performance-related clauses, the requirement to reduce the state's road trauma by 20% over a 5-year period, in accordance with the (then) government's traffic safety strategy target. This requirement was then reflected (passed down) in the key performance indicators of relevant subordinate staff. This clear statement of accountability had the benefit of conveying a sense of virtually unconditional support from the political level. It clearly implied that whatever support was needed from government would be pursued through the relevant minister.

Then there are the many and varied needs of organisations outside the government sector who face many of the same management challenges in supporting improved traffic safety performance while dealing with related impacts on other areas of their interest. They should be encouraged to play a much greater role.

The Safe System approach seeks to consolidate traffic safety improvements in recent decades and generate further improvement. In doing so, it explicitly adopts an approach focussed on outcomes and results,[135] it forces the nature of interventions to be reconsidered, and it relies on a systematic refocussing of institutional arrangements to implement those interventions.

It is suggested that countries with higher levels of traffic safety performance are in that space because (among other things) they have a robust system of governance for public decision making. It is likely that their systems display a readiness to

- Base decision making on evidence
- Prioritise actions on the basis of cost-effective reduction of serious harm
- Pursue unpopular initiatives that, based on the science and the recorded experience of others, will demonstrably improve traffic safety outcomes (that is, eschewing a solely populist policy development approach)
- Have in place the mechanics to ensure relatively noncorrupt (in a petty sense) enforcement agencies
- Treat all citizens without fear or favour, based on the rule of law

7 Serious Crashes Have Impacts Way Beyond Those Injured

We have, in the preceding chapters, painted a picture of where we are, how we got there, and where we might aim to go. As a motivator for action, we put a human face on serious injury in Chapter 2. Before turning to how we might start the new journey, we add a further dimension to the human face of trauma.

When we think of victims of road trauma, we tend to think of people like Jan and Noel, or Sam, or Abbey, people who (as we saw in Chapter 2) have been injured in crashes and who bear the physical and mental scars of their trauma. What we tend to overlook is that many more people are affected by each and every crash. In addition to family and friends, there are the unsung people who respond to road crashes, who are tasked with the job of telling parents that their child will never come home, or who try in vain to keep someone alive despite horrific injuries. For many emergency service personnel, despite it being part of the daily routine in their working life, they are deeply affected. Here are Richard and Kate's stories.

7.1 RICHARD'S STORY

Richard is a superintendent at a major metropolitan police service in Australia. These days, he is busy in the world of policing data, but it wasn't so long ago that he headed up his state's Major Collision Investigation Unit (MCIU). This unit was set up to respond to all culpable driving events in the state, those that have some aspect of criminality. Further roles for the unit are to investigate all crashes with three or more fatalities, and all crashes involving police personnel (regardless of fault) to provide an ethical standards oversight and to ensure the investigations are impartial. As the officer in charge at MCIU, Richard was tasked to personally attend all multiple fatalities and police-involved crashes. While in many ways the role had a high administrative burden, he was also in charge of monitoring the welfare and well-being of his staff on scene at major crashes.

The MCIU is a substantial unit with two offices, each comprising four investigation teams. These teams comprise a senior sergeant, three investigators, and an intelligence analyst. In addition, each office has three collision investigators and four motor mechanics. The response availability of the unit is 24 hours a day, 7 days a week, 52 weeks a year. All staff working in the MCIU are highly trained and highly skilled: all collision investigators have university degrees and, on top of their tertiary training, are sent overseas for specialist crash reconstruction training. The operating costs are high; in addition to the staffing costs (including considerable overtime), there is specialist equipment and custom-fitted vehicles to purchase and maintain.

It is a telling indictment that, despite our claims of the lowest road toll ever, MCIUs are required in each state in Australia, and similar units operate right across the Western world. Even more telling is that these units exist only to deal with the culpable and multiple-fatality cases, the miniscule tip of the iceberg among the huge number of simple mistakes that have the same result. The most thorough crash investigation system that we maintain is directed at prosecution, not prevention.

Richard's story does not start with his role at MCIU, however. It starts when he was a 19-year-old rookie police officer. It was a Sunday morning and he attended a call at a local park. A 2-year-old child had run out between two parked cars and been killed by an oncoming car. The child's mother and aunt had witnessed the child being run over; however, the father and uncle were in a different part of the park, and Richard was tasked with accompanying the child's mother to find the rest of the family and break the news. This was Richard's first experience with the devastation that follows road trauma, and to see a family torn apart in a split second is not something he can ever forget.

Three years later, Richard responded to a call where a 17-year-old apprentice had been out for a couple of drinks with his workmates, and while walking home was killed by a driver who was speeding and smoking cannabis. At that time, the police had no powers to take blood in order to test for alcohol or illicit substances. The hardest part of this case for Richard was reporting on the investigation at the coroner's court. Trying to explain to the parents why their teenage child had been killed and why nothing could be done about it. Richard had joined the police to help people, and provide justice for victims. Instead, he was left feeling helpless about the daily face of road trauma. Twenty years later, he found himself at the Major Collision Investigation Unit.

Working with major trauma every day is a challenge, although Richard muses it is probably not as shocking or difficult for the specialised MCIU team as it is for "ordinary" cops. Much of the work for the average police officer involves responding to burglaries, domestic disturbances, or broken letterboxes; responding to a fatal crash is an unusual, unexpected, and acutely stressful event. Conversely, the expectation for the MCIU team is that they will attend a fatal crash every single day. This reframes their expectations and can dampen the acuity of the stress; however, not everyone's mental health can withstand it. Even more problematic, the cumulative effect of repeated trauma response on individuals can be devastating.

In order to cope with the daily trauma that unfolds at MCIU, there is a great need within the team to share experiences. There is also considerable empathy among colleagues; it can be quickly recognised when someone is not travelling too well, or has been significantly affected by a particular crash.

Richard talks about the two main adverse impacts on the well-being of staff from working in such an environment. First, any one single crash can have a catastrophic effect. This might be a crash where the deceased is a child or a baby (maybe the person has his or her own kids and this is too close to home). Sometimes the victim is a teenager, or might look like a relative or friend. It may be that the scene is particularly challenging; there could be a lot of blood, or limbs strewn across the road.

Other times, the person looks very peaceful, as if they are asleep, with nothing to suggest that they will never take a breath again, and this can be really hard to get your head around. The problem is, for each person on the MCIU team, an acute stress trigger is different, and you never know when you turn up at a crash if this will be one that will haunt you for years to come.

The second adverse effect from working in major collisions is the cumulative effect of trauma. Everyone has a limit that they can be exposed to; there are only so many cases you can attend and so many faces of death you can look at before you reach that limit. Again, the problem is that everyone's limit is different, and you never know when you turn up at a crash if this is the one that will be the catalyst that pushes you over the edge.

It is not just the trauma of the crash scene itself that members of the MCIU have to deal with on a daily basis; it is also the devastation caused to the families where a loved one has died. The hardest part for Richard is breaking the news to a family, and people go through a range of emotions before they are even able to listen and comprehend the conversation. Sometimes, the families can't get themselves to a rational state after a crash. Which part do you focus on with the family when it is not clear what happened? Was it the road design, the weather, the speed? Families can obsess over these factors for years, but it will never change the outcome, and often they will never have a clear-cut answer.

Richard talks about one case where they were working with a mother who had lost her teenage son in a crash. This woman had extreme difficulty in coping with her loss, and would ring her investigator every day. She became suicidal, and unable to cope with her other children. This woman's anguish is not something Richard's team could fix, despite arranging for counselling and talking with her at length each day. That in itself is a heartbreaking job. Richard also recalls a situation where a mother who had lost her son in a crash was devastated that the person she felt was responsible hadn't been charged. This, of course, had nothing to do with the MCIU investigator; the decision was a legal process. However, the woman was so angry and grief-stricken she needed to lash out, and asked the investigator if he had children. When he replied that he did, she told him she hoped that one of his children would be killed so he would know what it felt like.

Despite the very real stresses that are part of the territory when working in major collisions, there are also aspects of the job that can be protective against some of the more negative stressors. One such aspect is the simple action of actually having a (very important) role to play at a crash scene. You don't just turn up and survey the carnage and take in all the stress and emotional tension around you; instead, you are very focussed on the particular job that you have to do, whether that be taking measurements, or inspecting the vehicle, or looking for debris on the roadside. Richard comments that in situations when extra team members turn out to a scene to see if there is any help needed, but don't have a specific role to play, this can often be when they are most negatively affected by a crash scene.

Actually, Richard notes that it is not only the police and other emergency services personnel that can suffer at a difficult road crash scene; members of the media who have turned out to cover the story can also find it very traumatic. Journalists have a tough job to do when covering crashes; they are usually pretty respectful of the deceased and their families, and in situations where the media do become "feral," it is simply a reflection of the stress (and distress) on the scene not being managed properly. In fact, members of the media even have dedicated groups to deal with the ensuing posttraumatic stress from witnessing the aftermath of major road crashes. We do forget that all these people are victims too.

Richard's years in cleaning up the mess of major road trauma leaves him frustrated by the whole state of affairs in road safety. To best articulate his frustrations, a notable example is a major rail crash that happened a few years ago. On a rural road, a semitrailer travelling at 100 km/h hit a busy passenger train. Richard attended this scene, where 11 people died and a further 23 were injured. Not only did the rail crash story make front-page Australian news for days on end, but the media made a huge issue of aboveground rail-level crossings (which is the norm on Australian roads). It took a truly catastrophic event to force a huge refocus on road design.

The problem for Richard is, how many people did he have to see die before *preventive* action was taken? As the scale of the crash amps up, the more people come to help; politicians and every conceivable emergency service turned up to see what they could do (and rightly so). However, we only seem to learn lessons when we have a huge, catastrophic, multiple-fatality event. What is it that makes these high-profile crashes worse than any other crash? The following week, Richard attended a single-vehicle fatal collision nearby. This only made a one-liner news story on page 7 of the local paper, and not one single politician or community leader turned up. What made the rail crash more worthy of attention than the next 11 fatal crashes that Richard attended? None of these 11 crashes he attended following the rail crash garnered the same type of media attention, or resulted in a moral outrage that focused our attention on ways of improving safety on our roads. As we observed in the introductory section of Chapter 2, we are inoculated with apathy by the steady drip, drip of everyday frequency. These everyday crashes are, quite simply, lessons not learnt.

While we do have bits and pieces of prevention efforts filtering through slowly, there really needs to be a whole system shake-up and a complete refocus on road safety. Unfortunately, it always seems to take a tragedy for the focus to shift, and for the families and friends of those that have already died, it is too little, too late. For Richard, his job is about trying to make sense of a crash and reconcile why someone is dead. But the other question is, why should anyone die at all?

7.2   KATE'S STORY

Today Kate is a paramedic, but she originally trained as a nurse and spent 7 years working in a specialist road trauma unit at a major Australian tertiary hospital. The particular road trauma unit where Kate worked was a busy place that looked after many incredibly sick patients. The unit incorporated a general ward area, a high dependency unit (HDU) and an intensive care unit (ICU). Kate started work as a graduate nurse in the ward, before progressing into the HDU and finally to ICU before retraining as a paramedic and joining the ambulance service.

Each of these wards had its own challenges: people with massive head injuries, people with permanent tracheotomies, and young people at the start of their lives now destined to live out their days in nursing homes. Looking after people in the road trauma unit is a difficult job. The nurses have to not only worry about caring for people who are very severely unwell, sometimes with no hope of recovery, but also build relationships with and care for the families who are distressed and grieving. All too often, these families are riding the rollercoaster of not knowing whether their loved one will wake up, and if he or she does, if he or she will ever walk or talk again, but at the same time are thrown into the quagmire of legal proceedings and police investigations and potentially coping with other bereaved families. It is this impact on the families that Kate often found most tragic to deal with, helping the people who are picking up the pieces of a life that will never be the same again.

Kate recalls one case in the road trauma unit that has remained with her. A young man, in his late teens, was brought in with massive head injuries. He also had multiple fractures, and needed to be intubated for a long time. He had no real hope of recovery. However, his family just could not accept this. They had an understandable, but completely unrealistic, hope that he would recover and return to his old self. After a long period of time in the road trauma unit, at the family's insistence he was transferred to a rehabilitation ward despite not being a suitable candidate for a rehabilitation setting. As he had so many medical problems relating to his injuries (such as temperature regulation), his body was unable to cope without significant medical intervention and he kept getting sent back to the road trauma unit where the problems relating to his significant head injuries could be dealt with better. However, the family could not accept that this was how his life was now going to be; he would not improve from this state. Kate recalls him moaning and making the most heartbreaking noises, with the pain he was suffering quite evident. And the pain was there for everyone, for his family, for his friends, and for the medical staff who tirelessly cared for him every day.

What Kate took away most of all from her experience in the road trauma unit is a sense that life is fickle; it can be switched on and it can be switched off in the blink of an eye. And for all the young people that were wheeled through those doors covered in tubes and wires, they will never realise their dreams or become the people they imagined.

As a paramedic, Kate probably responds to one major crash each year, and around three less severe crashes each week. Ambulances are also required at all fatal crashes to estimate time of death. Dispatching ambulances to road crashes is a very important, and stressful, job. It is also one that Kate's husband is responsible for. The ambulance service uses a computer-aided system to help find the closest ambulance; however, it can be quite difficult to determine the exact location of a crash when people call in on their mobiles or the actual crash site is on a freeway/motorway or rural location. Of course, major crashes are very time critical, and it is really important to get ambulances (with the right services on board, such as mobile intensive care) there in the fastest time possible to get the best outcome possible. Despite being computer assisted, human input is required to determine the level of urgency (for example, nonurgent or a code 1 lights and sirens). This job has very little margin for error, a kind of stress that few people would have to face on a daily basis in their working lives.

When a call comes in and Kate knows she has to respond to a major crash scene, she has an overwhelming sense of trepidation. Many things run through her mind: What steps will I take when I get there? How many patients will there be? Will the scene be easy to access? In some ways, though, the actual response is one of the less stressful parts of the process. Like Richard, Kate notes that during the actual response phase, the defined roles and responsibilities and having a job to do protects you from much of the heightened emotion and carnage that unfold as you arrive. It is actually the part *after*, when you think about what unfolded, that is the difficult bit. Luckily, most ambulances have teams of two on board, which provides an ongoing opportunity to debrief in the ambulance after a stressful job. Sometimes, though, more intensive help is needed. Part of Kate's husband's role is to identify jobs that have not gone so well; perhaps the patient died or there were some other highly stressful characteristics. Paramedics that attended these scenes are taken out of action for a short time and offered more formal debriefing with a support person, with counsellors or psychologists available to explore any deeper issues that this may provoke.

Kate recalls one road crash that she attended as a paramedic that she still finds very upsetting. She was training a young student on his first day in the ambulance service and was called to a road crash scene close by. They were the first responders on scene. Two cars had been drag racing on a major arterial road. One car had fled the scene, and the other, containing four young teenagers, was wrapped around a pole. When Kate arrived, the driver had got out the car and run away. The front seat passenger had also got out and run away, but had collapsed and was lying on the pavement a short distance away, bleeding heavily. In the rear, one young man was dead and the other was dying. For Kate, this moment was chaotic. She was the first on scene. She had several patients to attend to. She had a student ambulance officer she needed to ensure was coping. Despite Kate's best efforts, the young man in the back of the car died in her arms. This was a story with intense media interest, and the fallout in the papers played out for days with stories, and family angles and obituaries from friends. Every time Kate opened a paper or turned the TV on she would yet again see the face of the teenager that she put her arms around while he took his last breath. It is all too often forgotten that it is not just the families and friends that grieve these deaths, but also the people that care for the victims in their last moments. Kate will never forget this young man, or the place where he died, and in her life outside being a paramedic, she is reminded of this moment often as she lives close to the crash scene and has to drive past the spot twice a week.

Kate and her husband find it beneficial to talk about their experiences at home; they discuss advancements in procedural and treatment options, and they debrief on jobs they have found particularly hard. However, sometimes, their job comes home with them and it is very difficult to leave behind. Kate's husband was badly affected by one road crash that he attended, not long after their first child had been born. A young man had been killed at this particular crash, and there was a child seat in the back of the car. Although there was no child in the car at the time, he was struck by the fact that the child that could have been in that car was now left without a father. As a parent he had an enormous amount of empathy. This young guy was just going about his daily business on his way to work, and then the next minute he was gone. And that little baby would grow up without a dad. He had a lot of trouble sleeping for quite a while after this incident.

Kate muses that we, as humans, are eternal optimists. It takes a long time for a family to come to a true understanding that someone we love is gone, or that they are not going to get better. Treatment and recovery (or lack of recovery) after a road crash is a long, slow process that all too often does not have an end in sight. But as the wave of road trauma takes hold, it is not just the victim and his or her immediate family that is affected. Think of Kate and her family, or Richard and his family. Going to work each day brings the potential of someone dying on the roads, for Kate and her husband having to make life and death decisions, for this to be the crash that is the limit of trauma that Richard talks about, and stops them sleeping night after night.

Road trauma is not just about the injured people. It is not just about their families. It is not just about the cost. It is about the community, all the people that are directly and indirectly affected, the nurses, the doctors, the paramedics, the police, the tow truck drivers, the witnesses. In one way or another, we are all victims of road trauma.

8 Approaching Traffic Safety as Preventive Medicine

8.1 WIFM, FREEDOM OF CHOICE, AND THE DILEMMA OF THE COMMONS

In Western society, the major public health challenges of the twenty-first century will be dominated by lifestyle-related threats: obesity, abuse of alcohol and drugs, and trauma resulting from the way we use the roads.[52,140] Compounding this there is a growing culture of immediacy that pervades all aspects of our lives; everything has to be fast, now, and personally gratifying—interpersonal communication (cell phones, social media), modes of doing business (Internet marketing, online sales), IT processing speeds, minimum personal journey times, and so on.[141] Reconciling the demands for immediacy with safe road use and resolving the disconnect between the perception that driving behaviour is a matter of personal choice and the reality of interdependent social behaviour both require fundamental culture change at the whole-of-society level. How we might set about achieving such lasting change is the theme of this book.

The strength of the motivation to maximise immediate personal gains goes a long way toward explaining why we drive the way we do. The common acronym WIFM (what's in it for me) captures it beautifully. In two highly popular books— *Freakonomics*[142] in 2005 and *Superfreakonomics*[12] in 2009—economist Levitt and journalist Dubner presented a compelling case for incentives to be regarded as the cornerstones of modern life. They defined three types of incentive that people respond to—economic, social, and moral. For a smoker, an economic incentive to quit is an increase in the tax on tobacco, a social incentive is the increasingly widespread ban on smoking in public places, while a moral incentive is the praise received from peers for refraining from smoking. They argue that the incentive to cheat in sport is compelling because of the value placed upon winning. If you cheat to win, people understand your motive; if you cheat to lose, they despise you. They also argue that we must not substitute an economic imperative for a moral one. A traffic safety example is the use of fines for speeding; you pay the fine with indignation but feel no guilt over your behaviour. Keep this in mind as we later explore society's views about low-level speeding, especially when detected by cameras.

This is not to suggest that altruism does not exist or that most people do not comply with most social mores—the social and moral incentives are powerful. However, the evidence suggests that immediate personal gratification will often win when in competition with relatively weak moral imperatives.

Coupled with the importance Western society places upon individual choice— upon autonomy and self-reliance—we can begin to sense why we need to address the culture around road use rather than only focus on changing specific unsafe behaviours at the individual level, the mainstay for decades of our current armoury of behavioural traffic safety measures.[112,143] Addressing the level of public ignorance of the widespread impact of the road trauma problem and debunking the myth of blame as the principal culprit must be part of the solution.

Of course, not all Western societies place the same degree of importance on personal freedom. The United States is at one extreme and nations like Sweden are at the other, but the continuum is one where individual choice is highly valued across the spectrum. Each of us wants to be master of our own destiny. The $64 million question is to what degree? What are—or should be—the limits to our personal autonomy?

The beginnings of an answer can be found in the dilemma of the commons, first described in the 1830s.[60,144,145] In brief, a finite parcel of land in a village is made available as a commons for everyone to use to graze their animals. Each individual stands to gain personally by grazing an additional animal on the commons, and although each animal added decreases the amount of land available for all, the impact of this loss to the individual is small—in short, the direct personal benefit outweighs the direct personal loss. This is the dilemma of the commons—if everyone behaves only for personal gain, the commons will collapse.

A commons is said to exist when

- Ownership of a resource is held in common (roads are publicly owned)
- There is a large number of users with independent rights of access to the resource (the whole population comprises road users of one form or another)
- No one user can control the activities of other users (clearly true for road use)
- Total use (demand) exceeds supply (certainly true at times of traffic congestion)

The public road system can thus be viewed, in essence, as a commons. Let's consider some examples of how the dilemma of the commons pans out in everyday traffic situations.

Speed behaviour is an excellent example. There is a huge scientific literature that demonstrates unequivocally that fatalities and serious injuries from road crashes are reduced whenever the prevailing speeds are moderated across the traffic stream, even by small amounts (this literature is considered in detail in Chapter 9). An individual driver who chooses to speed typically receives immediate rewards—passing obstructing traffic, catching a green light, enjoying a thrill, and so on. Depending upon the extent of the speeding (the degree above the prevailing limit), the increase in crash risk to that individual may be quite small and is frequently not sensed as a risk at all. A large number of drivers all taking a small additional risk results in a disproportionate collective increase in risk across the traffic stream, and this is why it results in more and more serious crashes. Not surprisingly, large numbers of drivers exceed the designated speed limits relatively frequently. Consequently, most police forces apply a tolerance to their enforcement of speed limits, taking no action until speeding is in excess of some threshold (typically 10%) above the speed limit. This informal rule reinforces a widespread belief that the signed limit is indicative only. Clearly, no dilemma is felt at the individual level.

Truck driver behaviour is a second example. Commercial incentives exist to break safety regulations (overloading, speeding, driving hours), and while the immediate individual gains exceed the (potential) individual losses, the collective safety losses from these commercial incentives are directly measurable at the aggregate trauma level.

What we have in a commons is a tension between the desires of the individual and the needs of the community. Frequently, cries of "nanny state" are heard when governments propose to mandate behaviours to protect the population at large (such as seat belt wearing, helmet wearing by motorcyclists and bicyclists), to proscribe behaviours (such as driving with a blood alcohol above a certain level), or to stringently enforce existing laws (such as through the widespread use of speed cameras to moderate speeding). The question becomes, how should governments seek to resolve the tension?

Of course, if people only ever behaved for maximum immediate personal gain, we would have a dysfunctional law of the jungle society. On the other hand, a risk-free society in which government assumes all responsibility would require radical limits on individual freedom. Democratic governments are elected by citizens to provide what Jeremy Bentham described in the second half of the eighteenth century as the greatest happiness of the greatest number—his axiom of utilitarianism.[146] This axiom was later refined by economists to maximum total utility, and ultimately to assessment of options via the benefit-cost ratio. How much public policy is actually determined by an objective measure of the greatest good for the greatest number, and how much by effective lobbying by interest groups seeking personal advantage for their constituents? (See Box 8.1.) We argue that public policy on traffic safety issues where the collective gains are swamped by perceived immediate losses at the individual level explain our inability to make fundamental changes to the way we use our cars. This is dealt with at length in Chapter 9 on the special case of speeding.

The society with, arguably, the strongest culture of individualism is the United States. Sociologists suggest that self-reliance and individual autonomy with limited responsibility for the collective needs of the community are deeply rooted within American culture.[148] This is reflected in a distrust of government and a resistance to anything perceived as limiting individual rights.

BOX 8.1   SOCIETY AND THE COMMON GOOD

EXTRACT FROM AN EXCHANGE BETWEEN SIMON AND HIS PSYCHIATRIST IN ELLIOT PERLMAN'S NOVEL *SEVEN TYPES OF AMBIGUITY*

Simon: ... promoting the pursuit of the personal rather than the common good.
Psychiatrist: Never in western history has there been less emphasis on the common good. What did Margaret Thatcher say?
Simon: There's no such thing as society.

Source: Perlman, E., *Seven Types of Ambiguity*, Faber and Faber, London, 2003.

While seat belt wearing is now mandatory in most U.S. states, wearing rates are low by comparison with other Western motorised nations (Figure 8.1).[149] While they have been increasing since the mid-1990s, front-seat occupant (national average) wearing rates are only of the order of 85%—compared with mid to high 90s in most Western motorised nations. Rates across states range from a low of 73% in Massachusetts to a high of 97% in Washington. It is noteworthy that the rates are highest in those states with a so-called primary law—a law where a police officer can pull over a vehicle solely for an observed failure to wear a belt, not as an adjunct to another observed offence. Only 30 states, plus the District of Columbia, have primary laws.

Motorcycle helmet wearing laws have an even more chequered history, and helmet wearing is very low by Western motorised nation benchmarks (see Box 8.2).

In Florida, in 2000, riders over 20 years of age and with at least $10,000 of medical insurance were exempted from the law. The president of a Florida motorcyclist rights group was quoted as saying: "We want it left up to the individual. I only wear a helmet when it is cold or raining."[154] The age and insurance caveats are typical of a government wishing to appease a lobby group while still paying at least lip service to societal impacts. The choice to wear a helmet only when it is cold or raining reflects the importance of perceived personal benefit. In 1999, when helmet wearing was still mandatory, 22 unhelmeted riders died; by 2004 the number of fatally injured unhelmeted riders was around 250. The power of individual choice overriding community benefit is dramatic.

An evaluation study funded by the U.S. federal government reported huge decreases in helmet wearing and massive increases in deaths and serious head

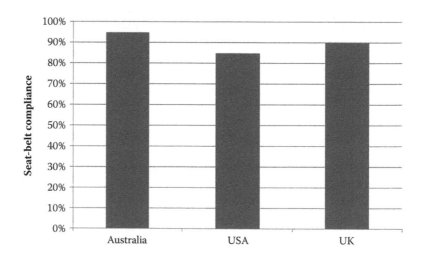

FIGURE 8.1 Seat belt wearing rates for drivers in Australia, the United States, and the UK. (Data derived from CARRS-Q, *State of the Road: Seat Belts*, Centre for Accident Research and Road Safety, Queensland, Australia, 2012; NHTSA, *Seat Belt Use in 2010—Overall Results*, National Highway Traffic Safety Administration, Washington, DC, 2010; ETSC, *Seat Belts and Child Restraints*, European Transport Safety Council, Brussels, Belgium, 2005.)

BOX 8.2   MOTORCYCLE AND BICYCLE HELMET
WEARING LAWS (UNITED STATES)

JULY 2011

- Laws requiring all motorcyclists to wear a helmet are in place in 20 states and the District of Columbia.
- Laws requiring only some motorcyclists to wear a helmet are in place in 27 states.
- There is no motorcycle helmet use law in three states (Illinois, Iowa, and New Hampshire).
- Some bicyclists are required by law to wear a helmet in 21 states and the District of Columbia.
- There is no bicycle helmet use law in 29 states.

The history of motorcycle helmet laws in the United States is characterized by change. In 1967, to increase motorcycle helmet use, the federal government required the states to enact helmet use laws in order to qualify for certain federal safety programs and highway construction funds. The federal incentive worked. By the early 1970s, almost all the states had universal motorcycle helmet laws. Michigan was the first state to repeal its law in 1968, beginning a pattern of repeal, reenactment, and amendment of motorcycle helmet laws. In 1976, states successfully lobbied Congress to stop the Department of Transportation from assessing financial penalties on states without helmet laws.

No state has a universal bicycle helmet law. Only 21 states and the District of Columbia have statewide bicycle helmet laws, and they apply only to young riders (often riders younger than 16). Local ordinances in a few other states require bicycle helmets for some or all riders.

Source: Insurance Institute for Highway Safety (IIHS), Highway Loss Data Institute, 2013, http://www.iihs.org/laws.

injuries as a result of the law changes in Florida, only a small proportion of which could be accounted for by the increased use of motorcycles over time.[155] A broader study revealed that motorcyclists involved in fatal crashes in states without a full wearing law were around 70% less likely to be wearing a helmet than motorcyclists in fatal crashes in states with such laws.[156]

Equally revealing is the case of the Pittsburgh Steelers quarterback of the time, who always wore his helmet when playing grid iron but not when riding his motorcycle.[154] Helmets and extensive body padding are mandatory in grid iron because body to body impacts are a feature of the game, whereas in motorcycling impacts are not anticipated and are considered avoidable by one's own skill level.

It is a common argument that individuals should be free to do what they want—to make their own judgements of the risks they are prepared to take—as long as their choices do not impact adversely on others. However, road use is very rarely, if ever, a solo risk-taking activity. Perhaps a lone driver on a road at a time with no other traffic and neither pedestrians nor roadside development could make a case, but even here, the costs of a crash impose financial costs on taxpayers and emotional costs on families and circles of friends and colleagues. In a 2011 survey of motorists across the United States, half reported having been personally involved in a serious crash or knowing a friend or relative who had been involved in a serious road crash.[59]

The American "live free or die" spirit seems to differ across states. Eric Wigglesworth compared fatality rates from various causes across U.S. states. He found that those states with the highest motor vehicle fatality rates also had the highest rates of firearm homicide, unintentional firearm deaths, drowning, and occupational deaths in the construction industry.[157] What this suggests is that cultural values at the local community level underpin public policy decisions and reinforces the case for addressing culture change as a key to future gains in safety. The Harvard story in Box 8.3 illustrates the ease with which scientific evidence can be dismissed where it is discordant with prevailing belief.

Australia, in contrast to the United States, has a reputation for legislating safety measures. It was the first country in the world with mandatory motorcycle helmet wearing, mandatory seat belt wearing, and mandatory random drug testing; among the first with mandatory bicyclist helmet wearing and mobile speed camera use; and has among the lowest proscribed blood alcohol levels.[143] High levels of enforcement, including widespread use of technology such as speed and red light cameras, supported by intense public education, are also characteristic of the Australian scene. Does this mean that Australians value collective responsibility more highly than

BOX 8.3   SUCCESS STORIES FROM THE HARVARD INJURY CONTROL RESEARCH CENTER

A wonderful book, *Mistakes Were Made (But Not by Me)*, uses the psychological concept of cognitive dissonance reduction to explain how people, often subconsciously, successfully weed out information that might contradict their beliefs.

A few weeks ago I gave a presentation about firearms to a half dozen congressional members of the Vermont House of Representatives. My talk focused on summarizing the studies relevant to the issue of a proposed Child Access Prevention law. After I finished the presentation, one of the members who was kind enough to listen, stated three times, and in public, "I don't care what the facts are. I don't care what the studies show" and then he went on to state his opinions.

—David Hemenway

Source: Hemenway, D., *While We Were Sleeping: Success Stories in Injury and Violence Prevention*, University of California Press, Berkeley, California, 2009.

Americans? Australians like to see themselves as laconic individuals with little respect for authority, so what explains the dramatic difference in traffic safety measures across the two nations, and what does this mean for the planning of culture change programs to change the ways we use our cars? (Section 4.2; see also Box 8.4).

BOX 8.4   GATEKEEPERS IN THE MEDIA AND IDEOLOGY STUNTING INNOVATION

Victoria, Australia, set out to lower the number of serious crashes in the early 2000s, with a particular focus on inappropriately high travel speed, a major factor in these crashes. Lower residential speed limits were introduced: 50 km/h as a default urban speed limit (down from 60 km/h, hardly leading edge speed policy!), and time-based 40 km/h speed zones outside schools and in strip shopping strips. In addition, tough speeding deterrence measures were introduced, which have been discussed elsewhere in this book. They included lower enforcement tolerances to those speeding above the limit, doubling of covert mobile speed camera hours, higher fines, more demerit points for speeding offences, and loss of licence for at least 6 months if more than 25 km/h over the relevant limit.

At a social function at the time, one of the media proprietors confided that they were concerned about the government's "grab for money" as being the reason for tougher enforcement. This played into the recurring theme from conservative news outlets about progressive governments having trouble balancing their books. It also builds on a narrower view of life held by many as being about money and not much else. This formed the basis for their vigorous resistance to the intervention. When advised by one of the authors that this view was disappointing, as the action by the government was based squarely on advice provided by the road safety agencies—who had been convinced by the evidence-based research showing that a substantial number of lives would be saved—the proprietor would not discuss the matter further.

This was a classic case of ideology refusing to be influenced by objective evidence through any discussion. While the issue has now died as a result of a conservative government being in place and due to the clearly demonstrated benefit of a major reduction in fatalities (up to 17% annually), this influence of ideology (from all sides of political thinking) continues to dog the free discussion of many new ideas in societies.

As J.M. Keynes is famously reported to have said: "When the facts change, sir, I change my mind. What do you do?"

8.2   PLACE FOR A PREVENTIVE MEDICINE APPROACH

Given that we depend so totally on our cars, that we view driving behaviour as a matter of personal choice, and that we—by and large each of us—blithely regard ourselves as above-average drivers, it is hardly surprising that we see death and serious injury on the roads as the fault of the aberrant few and not our personal problem.

The reporting by media and the institutionalised approach to determining fault, discussed at length in previous chapters, aids and abets this view.

However, as we have already shown, research at the University of Adelaide compellingly explodes the myth.[159] The researchers examined all crashes where a fatality had occurred in a given year and analysed the report from the (mandated) coronial inquest. They sorted the crashes into three categories—those where there was evidence of illegal, dangerous behaviour, those where there was minor illegal behaviour that did not appear causal, and those where there was no illegal behaviour. Approximately 40% of the fatal crashes involved illegal and dangerous causal behaviour, a sizeable minority but hardly in accord with the popular view. The university also operates an ongoing at-scene crash investigation unit. They examined all their data for nonfatal casualty crashes in the same year as the fatal crashes and sorted these according them to the same three criteria. Only 10% of these crashes involved illegal, dangerous causal behaviour.

In summary, the vast majority of nonfatal casualty crashes and a majority of fatal crashes do not involve aberrant behaviour—findings dramatically at odds with the prevailing public view. Nonetheless, as one of the leading lights of preventive medicine so eloquently put it: "The supposedly 'normal' majority needs to accept responsibility for its deviant minority—however loath it may be to do so."[130] Society seeks to distance itself from its deviants, but they are only the tail of the distribution. There is a continuum between the "sick" and the "normal." Hypertension, for example, is a continuum where the difference between "treat" and "don't treat" is large and grey, and the returns from a population-wide change in diet and exercise will be far greater than treating only those with blood pressure above a threshold level. Speeding is a great traffic example of this; most people exceed the speed limit sometimes, some much of the time but only by small margins, and only a few exceed the limit regularly by large margins. If everybody were to moderate their speeding behaviour, the collective returns would far exceed those gained by focussing effort only on the deviant extreme of speeders.

This leads to what is called the prevention paradox—it is a common irony that many people must take precautions to help the few.[130] Few of us have experienced severe load on our seat belts (no immediate personal gain), but because most of us wear belts all of the time, the minority who are involved in crashes are protected. In preventive medicine this is known as herd immunity. In the UK, Rose et al. estimated that for every 400 drivers belted, one life is saved. Forty years ago, 600 children had to be immunised against diphtheria for every life saved. If the risk reduction offered to an individual is small (as it is in both these examples), then the cost to the individual of that intervention must be correspondingly small.[130]

It is small—almost zero—for belt wearing. Note, however, that there was an elevated risk in the early days of mandatory seat belt wearing—poor initial designs required wearers to manually adjust their belts, but few people either knew about this need or knew how to ensure adjustment. Poorly adjusted belts created a new class of injuries known as the minced bowel syndrome. The appearance of these new injury patterns quickly led to a redesign that eliminated the need for manual adjustment.[78]

It is important to underline that governments had to mandate seat belt wearing in order to achieve herd immunity. For decades in Australia extensive public education had failed to bring voluntary wearing rates above about 40%. The rate jumped to well over 70% immediately after the law was passed, and then steadily rose to its present levels.[78] The law was the equivalent of mass immunisation.

Mass immunisation is an enormously effective public health measure and has saved more lives than any other medical intervention in the last century.[160] A number of diseases have been completely eradicated and others rendered very uncommon as a result of mass immunisation programs. Immunisation rates are now reasonably high—2010 figures reveal over 90% of Australian children were fully immunised against the key childhood diseases, somewhere between 85% and 97% in the UK, and more than 77% in the United States.[161–163] A favourable benefit-to-risk ratio has been well documented, and the WHO recommends 95% vaccination coverage as a national target.[164]

However, there has been a fall in immunisation rates since the 1980s. This has been attributed to the loss of trust in health professionals (similar to the loss of trust in police and governments found in traffic safety), and this refusal to vaccinate children is met with frustration by health and medical services. There are a number of arguments used in the immunisation debate that are circular—the debunked research linking the MMR vaccine with autism, the historical origins of some vaccines in aborted human foetal tissue, incorrect claims that current vaccines contain mercury, the myth that the diseases prevented are mild and self-limiting, and the argument that the parents' responsibility is to their own child and not others'.[165] In 1989, low vaccination rates facilitated a measles outbreak in the United States, which resulted in more than 55,000 cases of measles and 136 measles-associated deaths.[161] This is a topic that remains hotly debated online and shows the strength of feeling around individual versus collective rights.

The media must accept some responsibility for the standard of debates around public health issues. There is a tendency for the popular media—and particularly the "shock jocks"—to focus on populist themes, not on the facts. Stories of conflict sell, as we saw in Chapter 4. Traffic safety professionals must also accept a share of the blame for the lack of informed debate. We are poor communicators and, by and large, neither media-savvy nor accepting of our responsibility to raise the standard of public debate.

A population-level strategy is essential whenever a low level of risk is spread widely across a population. A 5% decrease in the average blood pressure of citizens has been predicted to lead to a 30% reduction in the number of strokes occurring in the population, compared with a reduction of only 15% if all cases of medically defined hypertension were successfully identified and treated. Similarly, a 10% decrease in average cholesterol levels is estimated to lead to a 20–30% decrease in premature heart disease.[160] A reduction of 4–5% in road crash fatalities is predicted to result from every drop of 1 km/h in average traffic speeds.[57] Public health prevention efforts, based on population strategies, are claimed to account for about 25 of the 30-year gain in life expectancy since the 1990s.[166] It is well and truly time that public health strategies were brought systematically to bear on the traffic safety problem.

The effect of fluoridation on dental health is well documented. Note that while it is also a population-level strategy, it does not require a population lifestyle change, unlike changing the incidence of hypertension, cholesterol, or speeding. It is a protective measure directly imposed by a government. The parallel in traffic safety is the application of design measures to eliminate death and serious injury from classes of crash, for example, the use of effective roadside barriers to eliminate collisions with fixed objects off the carriageway. The potential of design to impose population-wide change in behaviour—or behavioural outcomes—was extensively discussed in Chapter 6.

The alternative to a population-level strategy is the high-risk strategy where you treat only those at exceedingly high levels of risk, for example, those with very high blood pressure or those who speed at deviant levels. This is a popular approach in traffic safety—we frequently hear calls to focus on drunks and "hoons" and to leave us "normal" people alone! By blaming only the minorities we exonerate ourselves from responsibility. But as Rose points out: "If a high risk group is defined broadly then most of those included will not actually prove to have a problem, but if it is defined narrowly, it can contribute little towards reducing the total burden of disease. What is best for the selected individuals is worst for the community."[130]

This does not mean that we should ignore the deviants, far from it—we must continue to try to deter illegal behaviour and to rehabilitate recidivists or require them to drive only cars fitted with technology that precludes further offences and offers a degree of protection to the community. But it most definitely does mean that we must not regard these measures as the only weapon we have to achieve a higher level of traffic safety. Indeed, we must recognise that population-level strategies will provide far better returns.

This is a difficult thing for many to accept, that each of us must behave in ways that may require us to forgo an immediate benefit in order that unknowable others will benefit in the long run. Yet we already do this in some cases, mostly without understanding that that is what is happening. Consider the blood alcohol concentration (BAC) maximum permitted under traffic law. In most countries it is a very low level—0.05 in Australia, 0.02 in Sweden, and so on. The introduction of so-called per se laws goes back, in Victoria, to the 1960s.[106] Before that, the police had to produce evidence of impairment in order to obtain a drunk driving conviction, a notoriously difficult thing to do and a goldmine for lawyers defending clients. Gradually, research results accrued showing the effects of different blood alcohol concentrations on driving performance.

The most critical research used what is known as the case-control method. Drivers in a sample of real-life injury crashes were tested for blood alcohol concentration. Then, at the same times of day and days of week, and under the same weather conditions, many drivers were stopped at the sites of each of the case crashes and asked to provide a breath sample. The blood alcohol levels of the normal (no-crash) drivers were compared with those involved in casualty crashes to compute the risks of a casualty crash as a function of the blood alcohol level. The results from the United States and Australia are remarkably similar.[167,168] Setting the risk of a crash at zero blood alcohol level as 1, the risk at 0.05 BAC was 2, and it exponentially increased until the risk at 0.15 BAC was many orders of magnitude greater.

How does a government use data of this type to set a legal limit for blood alcohol concentration? Most states in the United States initially chose 0.10 BAC (several chose higher) on the grounds that the risk was at least 10 times higher than at zero; in short, the decision was to proscribe only (relatively) extreme behaviour. (The national standard in the United States is now 0.08 BAC.) Others (for example, Victoria, Australia) chose 0.05 BAC, being the point where the risk was double that of zero, arguing that a government had the right to intervene on behalf of its citizens when risk doubled. All other Australian states eventually lowered their 0.08 BAC limits to 0.05 in the interests of national uniformity, and the arguments of the time as to whether the national limit should be 0.05 or 0.08 BAC are instructive. The case for 0.08 was that in very few crashes where alcohol was present were the BAC levels between 0.05 and 0.08, the vast majority being over 0.10, and thus adopting the lower level would unreasonably constrain citizens. The case for 0.05, which eventually carried the day, was that as very few people can accurately estimate their alcohol level, conveying a message that almost no drinking before driving was acceptable provided the simplest and clearest message. The decision was made in favour of community benefit over individual loss.

It is important to note that relative risk was not, in itself, very important in this example of public policy decision making in Australia. As we have already seen, the risk of a casualty crash on any given trip for an individual driver is extremely low, and doubling that risk (by allowing a level of 0.05) still leaves the risk for that individual driver (at 0.05) at a quite small level. However, while a high blood alcohol level dramatically increases crash risk, very few drivers are at those high levels and the aggregate number of their crashes is relatively small. The benefit of a low level is the behavioural standard it prescribes for the vast majority, and the more acceptable intense enforcement becomes in the eyes of the community.

An interesting example of a socially acceptable application of a population-level strategy is the programs we impose on young, inexperienced drivers who have dramatically elevated crash risks, especially in the first 6 months after obtaining their driving licence.[169] We impose limits on the power/weight ratios of the cars they can drive, set the blood alcohol level at zero, prohibit all cell phone use while driving, apply passenger restrictions, and even apply nighttime curfews—each of which restricts road use by the population of young drivers despite the fact that the vast majority will not be involved in a casualty crash. Late-night curfews are an astonishing example of where a highly elevated crash risk holds sway in decision making over an incredibly small component of the totality of young driver casualty crashes. It is fascinating that the United States leads the way in population-level measures addressing the young driver problem—although many states retain among the lowest legal licensing ages! How can we reconcile this with maximising individual freedom? Perhaps the young cannot vote!

It is tempting to conclude that public policy making does not operate on consistent rules, arguably because our decision makers are not approaching the specific issues from any underlying theoretical perspective, such as preventive medicine. Sometimes we treat extreme risk but low incidence (night curfews); sometimes we go for true population targeting (low risk but high incidence). But even here we are inconsistent despite incontrovertible evidence of potential benefit. In Australia, a low blood alcohol concentration limit coupled with intense random breath testing is widely accepted within the community. But lower speed limits with intense zero tolerance enforcement, particularly via technology such as speed cameras, are highly controversial.[170]

Perhaps, at the end of the day, it is because normal means common or usual, not necessarily healthy. Rose pointed out that in the Pacific Islands fat is normal, and that in a rural Nigerian community rubbing cow dung into the umbilical stump of newborn infants was normal, despite 30% dying of tetanus.[130] In our society low-level speeding is normal. The formidable difficulty we face is convincing people that it is not healthy (for the community at large), though it is often advantageous to the individual at the time it happens. At present there are few, if any, incentives for individuals to change.

Public health deals with the collective interest, while conventional medicine deals with the individual interest. What are the objections to better preventive medicine? Ignorance, vested interests, and cost are the key ones. How, then, do we motivate people to change? A middle-age heavy smoker who gives it up only improves his longevity by a small amount. An obese person who wants to lose weight gets little benefit in the near term. As individuals we want benefits that are visible, imminent, and probable. So we work on smokers by promoting their self-esteem and the social approval that goes with giving it up. What might we do in road traffic to substitute for the lack of immediate direct rewards from, say, slowing down? But despite all the negative popular commentary, speed compliance has improved in most countries where intense enforcement (especially with cameras) operates. We need to measure what the public does, and not be overwhelmed by what the noisy minority say!

Advocates for population-level strategies are often faced with a criticism of being interfering social engineers. However, reverting back to the commons argument, change is justified where an individual's actions can be shown to have an adverse impact on others. Florida repealed its helmet wearing law for those with a minimum level of medical insurance, ostensibly to counter the argument that riders who crash hurt only themselves. Insurance coverage meant the community did not have to share the burden, at least not directly. It is difficult, in traffic, to find examples of where risk taking by an individual does not impose burdens on the collective. Road crash victims groups frequently demonstrate cases where tragedy has occurred entirely as a result of the actions of others. Where is the boundary between individual responsibility and collective responsibility? There can be little doubt that the boundary is culturally determined and varies quite dramatically across motorised nations!

We should, however, live in hope. The advent of seat belt wearing laws was preceded by huge public controversy, but after 40 years or so is just part of the driving landscape. The advent of per se laws for drunk driving, and for random breath testing as the primary means of enforcement, was similarly controversial and bitterly fought by segments of the alcohol industry, but is now an accepted part of the driving landscape. We can foresee a day when compliance with safe speed limits is a normal part of motoring life. Indeed, in cities like Melbourne, Victoria, some 99% of drivers passing speed camera sites are complying with the limit, but sadly, the media is still focussing on the number of speeding fines issued.[171]

8.3 INSTITUTIONS, VESTED INTERESTS, AND POLICY DECISION MAKING

Another complication is that the varied objectives of transport planning, road system design and construction, and road transport system operation are the responsibility of different institutions; thus, institutional competition comes into play—each institution is charged with maximising progress toward its own key objectives.

Captain Arthur Phillip developed a town plan for Sydney in the very early days of the new colony of New South Wales (the birthplace of European settlement in Australia in the late 1780s), but it did not get implemented for 20 years because the colony's primary needs were food and housing. It is fascinating to note, though, that the first road built was to the governor's (Phillip) mansion![172]

Vested interests always cry wolf and twist evidence in support. Catalytic converters, when first proposed as a mechanism to reduce vehicle pollutants, were objected to by the automobile industry with claims that they would cost $1,000 per unit. When introduced, they in fact cost about $100 each, and the health savings in the UK were estimated at over $5 billion per annum.[173]

How do we overcome the silo mentality associated with the separation of institutions from the objectives that appear to compete? Holistic thinking might lead us to a win-win approach rather than what appears now, as someone wins and someone loses. So long as the primary criterion is the dollar, safety is almost inevitably the loser.

What gets funded/approved/implemented tends to be chosen because of one or more of the following:

- It aligns with popular belief.
- It oils the squeaky wheel.
- It is the lowest-cost fix.
- It is the action most likely to produce results in the short term.

This seems to us to describe well why Australia's traffic safety efforts relied so heavily for so long on legislation/regulation, enforcement, and supporting public education and substantially underinvested in safe infrastructure and safe vehicles.[174] Without a political and institutional safety culture, there will not be stretch targets, and without stretch targets there will not be optimal policies and practices.

The lack of an overarching safety culture underpins how we think about risk and how we measure progress. As Lindblom and Woodhouse point out: "The ability of every contemporary democracy to probe social problems and policy options is systematically crippled."[175] The community and its politicians do not understand science at all well. They only hear the most eloquent advocates. When issues are complex, there are many vested interests, with each presenting the evidence from its perspective. Decision makers get confused by arguments between experts, especially when there is controversy about recommended action (such as speed enforcement methods).[99] In Chapter 9, addressing the great debate around the role of speeding in road trauma, we will see many more examples of conflicting research results publicly spruiked by respectable scientists.

Proponents often create unrealistic expectations of outcomes through what has become known as myth propagation. For example, when the disposable diaper manufacturers were under threat in the United States from environmentalists who mounted a case to ban them on the basis of the threat of waste to landfills, they hired a reputable consulting firm to research the issue and "proved" that resource usage was at least as high for reusable diapers. It was all done by mathematical modelling and, of course, turned on the assumptions underpinning the model, such as the water and power needs of washing diapers.[176] The point is not whether the research was flawed, but that the "evidence" developed was sufficient to overturn the draft federal legislation. Myth propagation is manna to a vested interest!

We bang on continuously about the need to base action on scientific evidence. Evidence is meant to be the sheet anchor for rational policy development, but it is often the handmaiden of the vested interest advocate. Facts, interpretations, and options flow from partisan groups most of the time. Scientists have to learn to communicate simply, and vested interest research needs to be exposed. Sadly, it is difficult to counter the results of vested interest research, as the technical issues are typically beyond the understanding of the public and politicians. We need to figure out how to better connect the ivory tower to the corridor of power.

In a delightful press article titled "Democracy for Dummies", a Melbourne journalist described the "post-truth environment of contemporary political culture" through a number of examples.[177] One was that all the candidates during the 2008 U.S. presidential campaign said something must be done about petrol prices, but no one was prepared to tell Americans (least of all the "good ol' boy" of Texas) that the pickup truck should be replaced by an electric Toyota. Similarly, the then opposition party in the state of Victoria, Australia, railed against speed cameras during their election campaign, sensing the public controversy of the time, but adopted a totally different stance when in government.

8.4   WHAT CAN WE LEARN FROM OCCUPATIONAL SAFETY?

We have, for a very long while, compartmentalised traffic safety as a problem largely for the transport community to solve. Road design, construction, and maintenance; traffic management; vehicle safety standards; vehicle registration and driver licensing; and the creation and amendment of traffic rules are, in almost all motorised nations, the institutional responsibility of transport agencies and the police. Compartmentalisation has constrained our thinking, our policies, and our practices in traffic safety.

While the previous section explored the relevance of public health and preventive medicine to addressing traffic safety, few health agencies have been directly and actively involved in the creation of traffic safety policy and practice. This is despite the fact, as we showed in Chapter 5, that a public health professional was appointed the inaugural head of the national traffic safety effort in the United States, and that he revolutionalised the way traffic safety interventions were conceptualised by bringing the public health concepts of host, agent, and environment to traffic safety thinking.[104] Despite a handful of pioneering cross-disciplinary individual contributions, institutional responsibility has stayed with transport agencies because traffic safety has been seen by governments as simply a by-product of personal mobility that needs to be managed. The failure to systematically engage with public health agencies has severely restricted systematic adoption of some of the most effective strategies available.

There is a glimmer of hope. The World Health Organisation (WHO) undertook a study of public health issues and forecast that road trauma would rapidly climb to a position of prominence globally as some of the most heavily populated nations motorise. Through the United Nations this has led to a declaration of a Decade of Action, which is being led by the WHO (see Chapter 3 for details). How health works with transport to give effect to the ambitions of the UN resolution remains to be seen.

Similarly, there is a lengthy history within industry describing how occupational safety was gradually brought under control, and yet the lessons learned here have not been studied systematically by the transport sector. It is tragic that our thinking continues to be dominated by the traditions within our institutional silos.

There have been four eras in thinking about industrial safety:[178]

- **Technical period:** When industrial developments were rapid, catastrophic technology failures occurred frequently and the focus was on technical improvement.
- **Human error period:** As machines became more reliable, the limitations of the operators came into focus and training took precedence.
- **Sociotechnical period:** The need for a good system interface between man and machine came to prominence, combining design and training
- **Safety culture period:** Currently the importance of the culture of an enterprise in overarching Safe System design and operation is paramount.

Exploring current thinking provides some fascinating parallels for traffic safety.

Safety remains driven by the simple principle of the complete elimination of technical breakdowns and human error.[124] As we have moved closer to this ideal in industry, the focus has shifted to group behaviour (organisational culture). Safety culture is usually defined as a set of safety-related attitudes, values, or assumptions shared between the members of an organisation. It requires commitment and cooperation from all levels. This implies a unity and integration within an organisation that must be led from the highest levels, and when this does not exist, the consequences can be catastrophic.

The *Challenger* space shuttle disaster is an example of how power can overrule culture. *Challenger* blew up 73 seconds after launch. The immediate cause was a failure of O-rings supposed to seal the joints of the booster rocket. This failure was due to a loss of elasticity in the unusually cold weather. The engineers at the contractors warned NASA, through their seniors, of this risk and suggested a delayed launch. NASA, under criticism for the costs of its space program, wanted to press ahead. NASA strongly criticised the contractor, whose senior management changed their minds and recommended the launch proceed, against the advice of their own engineers. The fact that the engineers had issued a warning was never communicated to the upper echelons of NASA.[179]

Similarly, Perrow, in a marvellous book entitled *Normal Accidents*, describes how Ford and Firestone hid data about dangerous tyres from the U.S. federal government because of top management's concern over profits. These are corporate examples of how immediate personal gain (or the avoidance of loss) can overrule broader societal impacts.[180]

There are many similar examples from outside the safety field of the impact of organisational culture on individual behaviour. The Baring Bank's financial catastrophe was initially painted as the actions of a single rogue trader, but later was confirmed as a product of the organisational culture in which the employee reward system fostered reckless risk taking. The phone hacking scandal at the newspaper *News of the World* was initially portrayed as the work of a rogue reporter, but it later became clear that it was the product of an organisational culture that put sales ahead of ethics.

There has been a lot of research around safety culture in industry. Five types of culture have been identified:[181]

- **Pathological:** Who cares about safety as long as we don't get caught?
- **Reactive:** We do a lot about safety every time there is an accident.
- **Calculative:** We have systems in place to manage all known hazards.
- **Proactive:** We try and anticipate safety hazards before they arise.
- **Generative:** Safety first is how we do business around here.

The research evidence is clear. Proactive and generative types of culture lead to the best industrial safety records. What type of culture can we see in our governments and transport institutions around traffic safety?

Generalising—as we must—governments in motorised nations tend to operative reactively; "things" are put in place only at times of perceived crisis. This is not surprising given all we have said about the way communities view traffic safety. The more we consider road trauma as the result of aberrant behaviour by a deviant minority, the less likely we are to call for systematic action that will almost certainly impose additional constraints upon our personal driving behaviour choices. Governments respond to their perception of community desires.

In Victoria, Australia, there have been four periods of major traffic safety countermeasure innovation in the past 40 years. Each was preceded by an apparent surge in death from road crashes (see Figure 8.2). The first crisis occurred when the annual total of road crash deaths burst through the 1,000 barrier in 1970. A daily newspaper ran a series on the crisis and public outrage was generated. Compulsory seat belt wearing laws were the immediate outcome, and the climate of crisis outweighed vocal concerns over personal restrictions. The impact was immediate and the crisis disappeared. Some 7 years later there was an apparent surge in deaths, the community having come to regard the lower absolute number of annual deaths as the (now) normal state of affairs. Random breath testing was the major, but not the only, innovation to arise from this crisis, and again, it was accepted despite its apparent imposition on individual freedom. The same pattern has occurred twice more since; each time the apparent surge in annual deaths was relative to a regularly lower annual aggregate that the community had come to accept as the norm.

While governments seem to operate mostly at the reactive level, the leading institutions, at least in Australia, operate at the calculative, if not the proactive, level. The adoption of Safe System principles and the development of scientifically based traffic

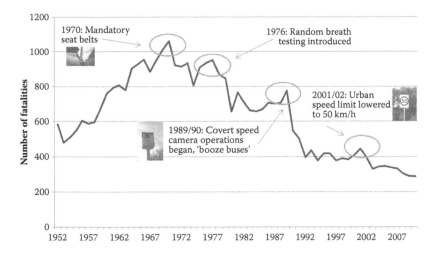

FIGURE 8.2 Number of fatalities and intervention introduction in Victoria, Australia. (Data derived from TAC, *Corporate Statistics: Online Crash Database*, 2013, http://www. tac.vic.gov.au.)

safety strategies attest to this. As the strategies are put to governments for adoption, political judgements of what is acceptable to the community are made and the strategies are adjusted accordingly. It is this normal democratic process—driven by the prevailing community view about the causes of serious crashes—that prevents the advancement to a generative approach to traffic safety. This is normal political behaviour, and we cannot expect politicians to change until the community stops regarding the traffic safety problem as one of individual behaviour and starts treating it as a societal problem.

One important foundation stone is to get the key institutions within road transport (road builders, traffic managers, vehicle makers, transport planners, etc.) to develop generative safety cultures internally. While Safe System thinking requires institutions such as road and traffic authorities to accept a level of accountability far greater than ever applied historically, this still falls fundamentally short of the generative safety cultures that can be found in the industrial field.

There are signs, in Australia, that this process has begun, but it is by no means complete. Governments and companies do not (generally) buy the safest vehicles in the classes they need operationally; transport agencies seek revenue from roadside advertising yet bemoan distraction as an issue; most police, at the officer level, do not enforce low-level speeding (though cameras do); there is a failure to hypothecate speed fine revenue to *additional* safety measures; there is a lack of action on vehicle advertising promoting speed and power; and many other examples besides.

In effect, those who plan, design, build, and operate the road transport system have no training in safety principles or strategies (Box 8.5). We have taken so long to progress because there was only a gradual shift from action based on experience, intuition, judgement, and tradition to action based on evidence, science, and technology. Most importantly, though, we lag because we have an inherently unsafe road

BOX 8.5 A SAFETY PRESENTATION TO ENGINEERS

An anecdote from one of the authors is in order. A middle level engineer challenged, during a presentation on safety back in 2007, why so much roadside barrier had been installed. His objection was that the repair costs were a drain on his budget as it was being constantly hit by vehicles! Others present quickly grasped the absurdity of the engineer's argument but to him the effect on his maintenance budget was the issue, not the saving of lives and injuries.

transport system and the resource and accountability implications of transformational change overwhelm planners.

In contrast, some large construction firms have implemented strong safety measures on their construction sites. All workers have swipe cards, and they have to swipe to gain entry to the site and again to leave. If a worker is observed practicing unsafely, his entry card is deactivated. When he turns up at work the next day, he is denied access and told to report to the office, where he is advised that his unsafe behaviour had been observed and he is not permitted to work that day (perhaps the following as well). The workers are not paid for this downtime. There are so many elements of cooperation in this scenario—the workers/supervisors/management taking responsibility for actions, incentives to act safely at all times, disincentives to slip on the standards—a clear example of the way in which safety needs active engagement at all levels.[183]

Ezra Hauer, writing on road and traffic engineering standards, talks of tribal customs and the logical flaw that says the proximal crash cause (typically a human error) must be the primary target for countermeasures. Standards are not based on either crash data or a real knowledge of human behaviour. Sadly, "things are made believable by uncritical repetition."[184] Design standards, guidelines, and practices are not founded empirically on safety knowledge. Vanderbilt adds: "The best laid plans of traffic engineers often run aground on the rocky shoals of human behaviour."[60]

Road crashes are the single largest cause of work-related disability and death, and yet are institutionally separated from mainstream occupational safety.[185] In Australia, for example, injury compensation and safety enforcement for workplaces are institutionally separate from transport. The taxi, bus, and truck industries, couriers, and safety for road workers are all matters for the transport agency, not the workplace safety agency, and compensation for injuries incurred on the journey to or from work are similarly separated. Silos are still alive and well!

Building on the success of generative safety cultures in industry is a second challenge. The potential for large public and private sector institutions—irrespective of their field of operation—to model safe driving cultures that are visible to the community at large may well be a critical step in catalysing a wider culture change. There is a significant opportunity for industry to take a leadership role, one that may become more widespread as the direct benefits to individual enterprises of a safety culture that extends beyond the immediate work environment are being increasingly documented.[186]

8.5   SAFE BEHAVIOUR, SAFETY CLIMATE, AND SAFETY CULTURE

Too often people say when an unsafe behaviour is lamented: "It's just part of our culture." Culture is a dynamic attribute, not a taken for granted state of play that can be used to explain behaviour, allowing us just to shrug our shoulders. Culture is never, in itself, an explanation. It is what needs to be explained if it is to be changed!

We have been quite successful at changing specific safety-related behaviours, at least in Australia. Protective behaviours such as seat belt wearing and helmet wearing for both motorcyclists and bicyclists are widespread.

Roadside sampling of blood alcohol levels across times of day and times of week suggest very high levels of compliance with our relatively low proscribed blood alcohol levels; typically, less than 1% of those tested are over the legal limit. Even targeted testing that focusses on times of day and days of week where higher incidence might be expected, reveal only 2–4%.[171] One must ask, however, whether we have changed the culture around drunk driving or whether we have simply changed the climate such that the probability of detection via intense random breath testing (RBT) is sufficiently high that all but those with serious alcohol problems and the social deviants are deterred?

Victoria introduced RBT in 1976, but the initial evaluation showed the legislation to be ineffective. A series of demonstration projects was conducted to explore the impact of enforcement levels, and it became clear that there was a threshold level of enforcement below which drivers did not consider the probability of apprehension sufficiently high to warrant changing their drinking and driving habits.[134] Once the intensity of enforcement was permanently boosted, the legislation became effective and the proportion of fatally injured drivers over the legal limit fell from one in two to one in five within 3 years.[134] How can we tell whether we have changed the culture around drunk driving or whether we have created a climate based solely on deterrence? If we truly believed the culture had changed, we would reduce the enforcement intensity, but the view, widely held among traffic police, is that the old patterns of behaviour would emerge fairly quickly. It is not an experiment that governments are seriously considering trying, despite the pressures on police resources!

While Victoria, in particular, has had considerable success in moderating speed behaviour,[187] it is arguable that there is not yet a lasting climate of deterrence around speeding, let alone a significant shift in the extant culture.

Culture change—a genuine lasting shift in normative behaviour—is unlikely to occur for specific behaviours such as speeding or drunk driving without a more widespread change concerning driving behaviour as a whole. The principles of safe road use need to become entrenched in societal behaviour generally. To improve safety, it is important to understand how the population thinks about safety—people need to make the connection between safe roads and safe road use and government strategies, values, and goals. This will not happen as long the existing widespread beliefs in misbehaviour and aberrant individuals as the primary issues to be addressed persist.

Changing culture and changing specific behaviours are different approaches. Culture change requires a greater emphasis on the environmental circumstances in which attitudes and values are formed, and the process by which behavioural patterns turn into attitudes and values. It also emphasises the influence of attitudes, values, and aspirations on behaviour, and thus how policy should respond accordingly. Behaviour change essentially rewards specific safe behaviours while discouraging specific unsafe behaviours. The two differ in where the intervention occurs: a top-down intervention early in the process at the culture level (culture change) or a bottom-up intervention at the exposure level (behaviour change). While (effective) programs to modify specific behaviours must continue, a broader effort to change underlying beliefs and attitudes around road use is also required.

The Safe System, explored in detail in Chapter 6, places a moral responsibility on those who design, build, and operate the road system to do their very best to ensure no one is killed or seriously injured. But this moral responsibility has not been codified—it is, in effect, a responsibility for *solving* the traffic safety problem.[188] This is the sense we should adopt when we use the Swedish term *Vision Zero*, as opposed to the traditional placing of total responsibility on the individual road user for his or her own fate. Elvik has correctly argued that zero is absurd if taken literally, as it would require a level of investment that would result in such a drastic redistribution of public monies that other ways people die or are disabled would be increased in frequency.[189] It needs to be seen as a forerunner of, or a catalyst for, a broader shift in the way we think about our cars and in the way we use our roads.

Let us end the chapter on an optimistic note—smoking has gone from cool and romantic (smoking in bed after sex) to unacceptable in almost all *public* settings. Has the legislation banning smoking in public buildings stemmed from evidence about the risk from passive smoking—the risk to others as well as to self—or is it a sign of a cultural shift? There has certainly been much tougher and much smarter advocacy around smoking. And it has taken a very long time, with major battles needing to be won against powerful vested interests. We must be prepared for a similarly lengthy process.

Similarly, recycling has taken on in a big way. While the decreasing availability of landfill was one driver, it was primarily the catalyst for efforts to modify widespread views around waste. Recycling in Australia is seen as a virtuous behaviour, and the social norm is one of a moral obligation to reduce waste.

It is often argued that the rapid rise of social media has led to a surge in a new form of bullying, and many schools are implementing antibullying programs. Research out of the United States suggests that the most effective programs are not that those target actual or potential bullies, but those that seek to engage bystanders who observe the bullying but simply remain passive observers.[190] Perhaps there is a lesson for traffic safety efforts in seeking to mobilise the "silent majority" to promote a new social norm around safe road use.

We have accepted the fluoridation of our water in the interests of our dental health, and security screening of our persons and our luggage at airports as reasonable interventions by government. We must seek to better understand what permits some interventions and not others.

The Neighbourhood Watch and Safety House schemes are further examples of cooperative behaviour. Our readiness to contribute to the victims of floods, fires, and the like demonstrates that the dilemma of the commons is resolvable at a societal level. Immediate gratification (selfishness?) on the roads is not an unalterable human trait, but rather a product of the traffic environment.

It should not be beyond us to achieve similar changes around our driving culture.

9 Speed Moderation: The Most Difficult Issue of All

9.1 WHY IS SPEED SO CRITICAL?

"Energy is the necessary and specific cause of injury."[191] We tend to forget this fundamental fact. No matter what the circumstance in which it occurs, human injury is, in the final analysis, the direct result of the impact of some form of energy upon the body that is beyond the limit of the body to tolerate. A burn is the result of too much thermal energy, a poisoning the result of too much chemical energy, while injury from a car crash or a fall is (almost always) the result of too much kinetic energy.

What do we know about kinetic energy, the (almost) universal direct cause of injury in a crash on the road? We have known for more than three centuries—since Isaac Newton described the laws of motion in the seventeenth century—that the amount of kinetic energy generated by a moving object is a function of *half* the mass (the weight) of the object multiplied by the *square* (multiplying the number by itself) of its *velocity* (speed), so speed is far and away the key determinant of the amount of the force unleashed in a crash.

Knowing this, much of the injury prevention effort in all fields of human endeavour has been directed at finding ways of minimising the amount of kinetic energy potential that exists in a given situation or ways of absorbing kinetic energy as it is unleashed, before it reaches the human body. For example, limiting the height of play equipment in a children's playground limits the maximum amount of kinetic energy that can be involved should a child fall (by limiting the speed a falling child can reach), while the installation of rubberised material underneath the equipment serves to absorb some of the kinetic energy released during a child's fall.

Here are just a few of the myriad traffic safety examples of attempts to *absorb* kinetic energy as it is released in a crash before it impacts the body:

- Inside the vehicle, manufacturers have, for example, padded instrument panels and removed all sharp surfaces, and have installed breakaway rear vision mirrors, collapsible steering columns, inertia reel seat belts, and airbags.
- Outside the vehicle, manufacturers have, for example, created crumple zones in the vehicle panels and ensured that engines are diverted beneath the passenger compartment.
- On the roadside, authorities have, for example, installed a wire-rope barrier that acts like a very strong (and nonrebounding) elastic band, and have created "slip base" utility poles designed to break away when struck.
- For bicyclists and motorcyclists manufacturers have designed helmets to absorb energy in the inevitable "second collision" when the rider falls to the pavement after a collision and experiences a head impact.

It is much more difficult to provide examples of attempts to limit the maximum amount of potential kinetic energy within the road transport system. Potential kinetic energy is a function of mass and velocity, so vehicle mass and speed are the sole determinants of potential kinetic energy. Antilock brakes and skid-resistant pavement surfaces are both designed to prevent skidding and so increase the reduction in potential kinetic energy that occurs progressively during emergency braking.

There have also been many attempts to reduce the likelihood that there will be a collision involving large amounts of kinetic energy. Obviously, if there is no collision, kinetic energy remains potential and therefore benign. Here are a few examples:

- Inside the vehicle there are antilock brakes and electronic stability control, designed to try and keep dangerous situations from becoming catastrophic.
- On the roads, authorities use signs and signals to control access to intersections, thereby minimising conflict, and use roundabouts to both reduce speed on the approach to an intersection and make the collision angles relatively shallow should a crash occur.
- Authorities try to avoid placing solid objects such as utility poles or trees close to the running lanes or at locations such as sharp curves.
- Training to raise skills, both in avoidance manoeuvres and in defensive techniques to decrease the probability of having to initiate an avoidance manoeuvre, is also employed widely to reduce crash frequency.

Setting speed limits across differing road and traffic conditions—and achieving widespread compliance with them—remains the fundamental tool available for managing the kinetic energy potential within our road transport system. After all, kinetic energy is half the mass times the *square* of the speed. In a safe system, speed limits will be matched to the level of protection from kinetic energy exchange offered by the design of our vehicles, roads and roadsides, and traffic flows. As we shall explore later in this chapter, that is not the case at present. First, though, we need to move past the theory of kinetic energy and its management to the scientific evidence concerning the role of speed in not just the severity of injury when a crash occurs, but also in the likelihood that a casualty crash will occur.

9.2   SPEED AND CRASH LIKELIHOOD

Speed itself is not the culprit—the problem is coming to a sudden stop! Some version of this trite statement is often heard in public debates around speed moderation. While Newtonian physics is vaguely understood, the key belief is that kinetic energy is benign unless and until it is released in a crash. It falls to zero when we stop at the end of our uneventful journey, and the vast majority of us make uneventful journeys every day of our lives. Our speed behaviour may involve levels of kinetic energy potential well above tolerable limits should a crash occur, but we believe we are in control and that a crash will not occur. We understand that a high-speed crash is a bad thing, but we do not accept that speeding increases the risk of a crash occurring. Many attempts have been made to assess scientifically whether, and if so to what extent, travel speed influences the likelihood that a casualty crash will occur.

The most commonly cited, and the most hotly debated, evidence is the frequency with which "speeding" is identified as a principal cause of crashes.[99] In most Western motorised countries, the incidence of speeding as a major cause has been reported, usually from police investigations of serious crashes, as being between 30% and 40% of all serious casualty crashes. While this figure is widely quoted in public debates, we are going to set it aside, as it is not sufficiently robust in a scientific sense. When police investigate a serious crash, they are actively seeking evidence of law breaking, and while their decisions as to whether or not the vehicle was exceeding the limit are usually accurate, it does not follow automatically that speed was a principal cause of the crash or, rather, a major contributor to the crash severity. This is the argument used by the vocal opponents of the evidence, and it has merit.[99]

There are three classes of scientifically robust evidence that collectively establish that higher travel speeds do result in a greater number of casualty crashes:

1. The case-control method. In this method scientists match their cases with controls and examine the differences. The University of Adelaide has conducted two comprehensive studies in South Australia, one in urban Adelaide and a later one on rural roads.[192,193] Only the latter study is described in detail here, as the conclusions to be drawn from the results of both are the same.

   As part of its ongoing research into crash causes and consequences, the university maintains an in-depth crash investigation team that attends a crash scene as soon as possible after receiving notification of a serious casualty crash, typically from an ambulance call. The team collects a vast array of data and, amongst other things, uses computer modelling to estimate the precrash speed of the vehicle.

   In this particular study the team examined crashes on roads on the fringes of Adelaide where speed limits were 80 km/h or greater. To qualify as a case, the vehicle had to be a passenger vehicle in which at least one person was seriously injured, it had to have been travelling at free speed (that is at the choice of the driver and not constrained by other traffic), the blood alcohol level of the driver had to be zero, and there had to be no evidence of a suicide attempt, or of a medical cause (such as a heart attack), or of an illegal manoeuvre, such as running a stop sign. From 167 crashes attended, 83 cases met all these criteria. In short, these were cases where there were good data to enable a reliable precrash speed estimate and to enable most other possible behavioural causal factors to be excluded.

   For each case crash, the team then obtained 10 controls (830 in total). They went to each case crash location at the same time of day on the same day of week and under the same weather conditions as the case crash and measured the free speed of 10 vehicles travelling in the same direction as the crash vehicle. They stopped these drivers farther down the road and sought breath samples to eliminate any alcohol influence.

   The average (measured) free speeds of the controls were compared with the (computer-estimated) precrash speeds of the cases to compute crash risk curves using logistic regression modelling—in effect, comparing the speed distribution of the noncrash vehicles with the speed distribution of the crash cases. The results suggested that an increase of 10 km/h in average travel speed resulted in a doubling of the risk of a serious casualty crash, an increase of 15 km/h increased risk 4-fold, and a 25 km/h increase in average speed increased crash risk 10-fold (Figure 9.1).

As one might expect, the higher the speed above average, the greater the crash risk, but the critical finding is that relatively small increases in speed from the average doubled the risk, a finding we consider again later in examining the impact of enforcement tolerances.

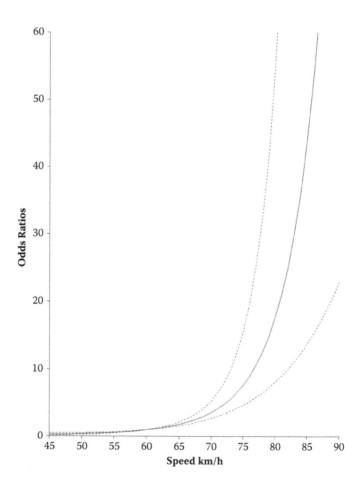

FIGURE 9.1   Crash risk curve at 60 km/h. (Reproduced with permission from Kloeden, C. et al., *Travelling Speed and the Risk of Crash Involvement: Findings*, Volume 1, Federal Office of Road Safety, Canberra, Australia, 1997.)

While the case-control method is sound, it is not infallible: in the field of traffic safety research we are trying to estimate real-world outcomes without the experimental control that comes in a laboratory, and scientists have to compromise as best they can. Not surprisingly, the University of Adelaide research was challenged because its findings ran contrary to popular belief. An Australian motoring club commissioned a traffic safety specialist from the United States, with expertise in mathematical and statistical techniques such as had been used in the Adelaide study, to conduct an independent review of the study and its findings. That review, while identifying some technical weaknesses, concluded that the approach was fundamentally sound and that the conclusion of an exponential relationship between speed above average and crash risk could not be denied.[195]

2. The effect of speed limit changes on crash frequencies. Numerous studies across many countries have evaluated the road crash injury outcomes of changes in speed limits, both urban and rural.

In the United States, a natural experiment inadvertently occurred. In 1973 the speed limit was reduced from 65 to 55 mph—105 km/h down to 90 km/h—on all interstate highways as a fuel-saving measure in response to a global oil crisis. The Transportation Research Board estimated that $2 billion in fuel savings accrued, that between 2,000 and 4,000 lives were saved, and that there was an increase of about one billion hours of total driving. In summary: fuel saved, lives saved, time lost.[196]

At the end of the oil crisis the lower speed limit was retained for quite some time as a safety measure. However, in response to lobbying based upon the estimated large increases in total driving hours, the limit that could be posted reverted to 65 mph in 1987, and by 1995 about 95% of the interstate network was back to a 65 mph limit, *and fatalities had increased*. In summary: fuel lost, lives lost, time saved.[197]

In 1995, the power of the federal government to set speed limits within state borders was successfully challenged, and each state became free to set the speed limit on those sections of interstate highway within its boundaries. Within a year, 23 states had raised their limits to 70 or 75 mph (120 km/h). Analyses show substantial increases in fatalities in these states, with interpretation needing to be qualified somewhat by regional differences in population, topography, and traffic flow and mix between states that stayed with 65, dropped to 55, or rose to 70 or higher.[197]

A similar natural experiment occurred in Victoria, Australia. The rural road speed limit was lifted from 100 km/h to 110 km/h on dual-carriageway roads in the mid-1980s, and an increase of around 20% in serious casualty crashes was observed. In response, the limit was reduced to 100 km/h again, and the number of casualty crashes fell by roughly the same proportion.[198] In every individual study seeking to compare crash and injury outcomes before and after changes in speed limits, methodological complications arise and no one result can be considered definitive. Because changes occur in the real world, not in the laboratory, the link between cause and effect can be clouded. Nevertheless, confidence increases if many studies report the same finding.

Rune Elvik, one of the most respected scientists in the field and, at the time, (joint) editor in chief of the international journal *Accident Analysis and Prevention*, conducted what is called a meta-analysis of all studies of this type that were reported in the scientific literature. In a meta-analysis, the first step is to review the scientific soundness of each individual study of the same issue and eliminate those judged to have fatal methodological flaws. Elvik identified 98 sound studies of the effects of speed limit changes in the scientific literature and, after analysing the collective results, concluded that there is a clear causal link between raises in speed limits and increases in crash casualties—and the converse.[199]

As an example of the converse, evaluations of the effect of reducing the speed limit from 60 km/h to 50 km/h in urban residential streets and selected collector roads in both Melbourne and Brisbane have, independently, shown reductions in the frequency of casualty crashes.[200]

Increases in speed limits increase travel speeds and, in turn, increase the frequency of casualties. There can be no doubt that small shifts in speed distributions (higher average and 85th percentile speeds), especially in urban areas, result in disproportionately larger increases in casualties.[201,202]

3. Correlational research. The third common method is to look for associations (statistical correlations) between different speed distributions and crash records. While correlations do not prove causation, the results can be informative.

One of the best studies of this type was carried out in England about a decade ago.[203] The authors were interested in the effect of different design standards of rural road, all with the same speed limit, on crash records. They also measured average travel speeds on the roads they examined. They found, as expected, that higher standard roads had fewer casualty crashes but, importantly for our purpose here, also found that for each standard of road, higher average speeds were associated with worse crash history records.

The inescapable conclusion from the weight of scientific evidence is that the higher the (potential) kinetic energy in the road transport system—the higher the travel speeds—the greater the frequency of serious casualties. Speed is indeed the culprit, not just the sudden stop.

9.3 WHY IS KINETIC ENERGY SO IMPORTANT?

Having established that higher travel speeds lead to more casualty crashes, it is important to understand why. There are several factors:

- Human decision making during driving involves perceiving continuously what is happening in the environment, interpreting the several signals constantly being received, deciding upon appropriate action, and implementing the chosen action. The time it takes to receive, process, and decide is called the *reaction time*. Individuals have different reaction times, depending upon such things as age, health, and experience with the task at hand. We all also have internal differences over time due to our current alertness level, the existence of distractions, and the like. In an emergency, we cannot, of course, begin to apply the brakes or execute an avoidance manoeuvre until the reaction time has elapsed, until we have sensed the danger and decided what to do. Reaction time is not related to speed of travel, but the distance travelled certainly is. If a typical reaction time is, say, 1 second, at 40 km/h we will have travelled about 11 metres, at 60 about 17 metres, at 80 about 22 metres, and at 100 about 28 metres before we begin to execute whatever emergency action we have decided upon.
- The stopping distance is also directly related to speed of travel, as well as to factors such as the state of the vehicle's brakes, the friction level of the road surface, and so on. Figure 9.2 shows total stopping distances for a range of starting speeds under assumptions of a reaction time of 1.2 seconds (a typical result from studies of unalerted humans in decision-making experiments), a modern vehicle with brakes in as-new condition, and a dry pavement with a coefficient of friction of 0.8 (a typical result derived from at-scene crash investigations in Australia).
- The higher the speed, the more complex the information processing task (for a given environment), as the sensory inputs are changing more rapidly and any distraction sources have their effects magnified.

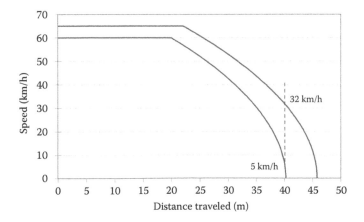

FIGURE 9.2 Stopping distance at 60 km/h. (Courtesy of Logan, D., and Corben, B., Speed Risk Curves: Unpublished Road Safety Modelling, Monash University Accident Research Centre, Melbourne, Australia, 2013.)

- The faster an individual is travelling relative to the prevailing traffic stream, the more likely that other drivers will misjudge things like closing speed, likelihood of conflict, and so on, and the higher the likelihood of a loss of control should an emergency arise.

9.4   EXTANT SPEED LIMITS AND CURRENT LEVELS OF PROTECTION

The management of kinetic energy exchange to levels that the human body can tolerate is a fundamental of injury prevention. As we saw in Section 6.2, there are five pillars of the Safe System. Setting aside the pillar *better management*, the other four are *safe vehicles*, *safe roads and roadsides*, *safe road users*, and *safe speeds*. Each is intimately connected to the others. Let us look at the extent to which the levels of protection provided by current vehicles, roads and roadsides, and road users match typical speed limits and commonly measured travel speeds.

Perhaps, after all, speed is not the culprit. It seems that the real culprit is that we have failed to match all the elements within the Safe System model. Safe speeds are speeds that are aligned with the available levels of protection, something we clearly do not do well enough.

9.5   SAFE VEHICLES AND SAFE SPEEDS

The crashworthiness of vehicles has improved dramatically over the past four decades or so. Most of the initial effort was directed at making crashes survivable, and much has been achieved. It has been estimated that a fairly constant percentage of the steady decline in fatalities in most Western motorised nations is directly attributable to these achievements.[131]

Nevertheless, the levels of protection against serious injury provided by modern vehicles are generally less than what is needed given permitted travel speeds. We have not matched speed limits to the levels of design protection that vehicle manufacturers have been able to achieve.

The risk of death rises rapidly with impact speed for specific crash types, as illustrated in Figure 9.3.

The outcomes of fatality risk analysis for these key crash types has led to the target speed ranges set out in Table 9.1, being referenced in public documents prepared in the Netherlands, the OECD, Victoria, and Western Australia. These target ranges of speeds would take some time to implement, and their implementation is dependent upon a level of community understanding and acceptance that does not currently exist. Practical early steps are available in the short to medium term, however, to demonstrate the substantial safety and wider community benefits of a range of speed management interventions.

These suggested limits are based on data showing the probability of a fatality as a function of impact speed. The low limit for pedestrian safety is not surprising since a pedestrian lacks any protection. There are promising developments in pedestrian-friendly vehicle frontal designs that address the issue of pedestrians being run under

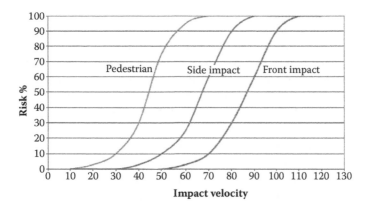

FIGURE 9.3 Risk of being killed as a function of impact speed. We are aware of ongoing scientific debate about the exact nature of these curves, particularly the pedestrian curve. However, these curves are considered a valuable means of improving community understanding about fatality risk versus speed and are the best available general representation of that risk, especially for older pedestrians, available at present. (Adapted with permission from Wramborg, P., Fatality Risk for Three Major Crash Types at Different Speed, in *Towards Zero: Ambitious Road Safety Targets and the Safe System Approach*, ed. Organisation for Economic Cooperation and Development (OECD), Transport Research Centre, Paris, France, 2005.)

TABLE 9.1
Safe Impact Speeds for Different Situations

Road and Section Types Combined with Road Users	Target Safe System Speed
Roads and sections used by cars and vulnerable users	30 km/h
Intersections with possible side-on conflicts between cars	50 km/h
Roads with possible frontal conflicts between cars	70 km/h
Roads with no possible frontal or side-on conflicts between vehicles and no vulnerable road users present	≥100 km/h

Source: OECD, Speed Management, in *European Conference of Ministers of Transport*, Paris, France, 2006.

rather than run over, but it is fair to say that we are a long way from solving this difficult design problem.

Similarly, side impact protection is quite limited; the very small space between a vehicle occupant and side intrusion severely limits the vehicle designers' options for absorbing kinetic energy before it impacts on the occupant. It is worth noting that the safety design rule for crash testing for side impact occupant protection requires a test at 50 km/h. Actual crashes at intersections on higher-speed roads often involve impact speeds well above this level.

We applaud the advances in occupant protection that the automobile industry has achieved, while at the same time noting that real-life speeds are well above what designers have been able to provide for. In motor racing, very high levels of protection for drivers are achievable, and these features (roll cages, helmets, full harness seat belts, etc.) could be transferred to the vehicle marketplace if that market was willing to face both the cost and the aesthetics. A separate question entirely would be how drivers might drive if they had race-level protection against extremely high-speed crashes!

The second plank of protection offered by vehicle design is the range of measures to reduce the probability of a crash occurring. The focus of vehicle designers has shifted markedly from crashworthiness—where the general consensus seems to be that most of what is practical has been achieved—to crash prevention. Electronic stability control significantly reduces the likelihood that an incipient loss of control event will result in a run-off-road or rollover crash.[207] Intelligent cruise control, pedestrian detection devices, lane departure warning systems, fatigue detection devices, alcohol ignition interlocks, and intelligent speed assist devices are all either promising or (close to) proven devices for assisting in crash prevention. Unfortunately, most are either fitted only for convicted repeat offenders (alcohol ignition interlocks) or offered only as options on the top-of-the-range models. The potential of the current design push is enormous, but we are a decade or more away from the technology permeating the bulk of the vehicle fleet, so we must adopt strategies to deal with the here and now.

9.6 SAFE ROADS, ROADSIDES, AND SAFE SPEEDS

The most common serious casualty crash on Australian rural roads is a run-off-road-into-a-fixed-object crash. Providing adequate roadside safety protection ought therefore to be a key priority, with the options ranging from raised pavement markings (to alert a driver through noise and vibration of an imminent event), through clearing roadsides of rigid objects (to provide recovery room), to substituting a collision with a rigid object with a collision with a "forgiving" roadside barrier. Arguably, however, rural roadside protection practice is not the key priority that it should be.

There is enormous potential to design and provide roads and roadsides that provide sufficient protection at extant speeds. A safe road, designed to sustain speeds of 100 km/h (or greater), can be profiled as a divided road (eliminating head-on crashes), with no intersections at grade (eliminating intersection crashes) and with either clear roadsides or roadsides protected by an appropriate energy-absorbing barrier (eliminating serious run-off-road crashes).

In many states of Australia, two-lane, two-way rural roads with gravel shoulders and tree-lined roadsides have speed limits of 100 or 110 km/h, clearly way above the designed level of protection. Australia has a large geographic area and a small population, which puts a massive strain on road expenditure. We constantly hear governments point to the lack of funding as a barrier to improving road and roadside safety. However, there are two ways forward: gradually improve the level of protection on the roads with the highest volumes and reduce speed limits to better match levels of protection on the remainder of the network. We shall turn shortly to considering why the latter is so hard to achieve.

A large proportion of all casualty crashes, the dominant proportion in urban areas, occur at intersections despite the provision of signs and signals to reduce conflicts. Many of these crashes involve vehicles from opposing roads or turning vehicles, and thus side impacts are typically frequent, yet intersection approach speed limits are the same as the prevailing limits on "uninterrupted" sections and well above the limits of protection offered by the vehicle. The introduction of intersection approach speed limits for major intersections is rarely considered and even more rarely implemented (Box 9.1).

9.7 SAFE ROAD USERS AND SAFE SPEEDS

We have already shown that our common speed limits are too high for the levels of protection offered through our current vehicle, road, and traffic engineering standards and practices. This is compounded by the fact that speed limits are honoured more in the breach than the observance, and that attempts to use intense enforcement to achieve compliance, especially through automated technology such as offence detection by cameras, are intensely controversial and many governments shy away from such an approach. Appealing to the public through education and training to moderate speed behaviour has little impact, not at all surprising when we consider the social context in which speed behaviour is determined.

9.8 HOW, THEN, DO WE SET SPEED LIMITS?

The management of kinetic energy has been, for the greater part of motorised history, a minor consideration in the setting of speed limits. At a macrolevel, urban streets where traffic is dense and pedestrian and two-wheeled road users are common have had lower limits than rural roads, but the limits have not been set to match levels of available protection as we have just seen. The primary consideration has been the facilitation of mobility, deemed to be, as we saw in Chapter 4, the minimisation of journey time.

In the 1920s, as motoring became more widespread in the United Kingdom, speed limits were abolished, except in densely populated areas, and were not reintroduced until 1934.[99] The rationale was that the police resented enforcing unpopular laws and were not supported by the courts when they did. One might also speculate that only the richer folk had motor vehicles in the 1920s!

In Australia, a general urban speed limit of 30 mph came into effect in the mid-1930s. In 1964 the limit was raised to 35 mph on the grounds that advances in vehicles and roads meant that the mobility benefits of the higher limit could be realised safely (though the lack of change in human capacity to withstand kinetic energy—above a threshold level—had clearly not changed). In 1974, upon metrication, the limit was set at 60 km/h rather than the more obvious conversion option of 55 km/h (30 mph = 56 km/h). As we have just seen, impact speeds of vehicles into pedestrians at above about 40 km/h greatly increase the risk of fatality, and McLean and

**BOX 9.1   THE SAFETY STORY OF A
LOCAL SHIRE IN AUSTRALIA**

The shire has an active road safety approach and sought to improve the safety of its rural and urban road network in the mid-2000s.

An independent consultant assessed the crash risk of its rural network and a series of practicable minor safety works were devised and implemented to reduce that risk (roadside barrier, shoulder sealing, etc.).

However, further assessment of crash risk after these works were completed indicated that it would be necessary to reduce some travel speeds to make some roads acceptably safe (from 100 km/h to 80 km/h).

Many of these rural roads are of a winding nature with narrow pavements hemmed in by vegetation, which could not be removed for environmental reasons.

At the same time, the shire wished to reduce speed limits on its residential streets from 50 km/h to 40 km/h for reasons of safety and amenity. It proposed a trial area.

The shire has a duty of care for the inherent safety of its local road network. While it had approached the state government road authority about improvements and reduced speed limits on some of the state roads, it was now seeking to reduce the risk travellers faced on its own local road network.

The shire faced resistance for a while from the state road authority, which has the sole authority to set speed limits on all roads in the state. This poses an interesting dilemma for local authorities in terms of responsibility for safety on the local roads.

In due course, the state road authority moved to support the shire in its desire to reduce speed limits on some roads and in the residential trial area. At the same time, the council staff were endeavouring to convince an

Anderson estimated this metrication decision has cost thousands of pedestrians their lives in Australia.[208] It is worth noting that the 60 km/h limit is high by international standards—50 km/h or lower is common through Europe and 30 mph (55 km/h) is typical in the United States.

For a long while, speed limits were set for each road class using the 85th percentile free speed measured for travel on those road classes.[117] The 85th percentile is the speed at or below which 85% of the vehicles are travelling, and the assumptions underlying its adoption were twofold: first, that drivers have a pretty good feel for what is both a desirable and a safe speed for the circumstances, and second, that by providing for the wishes of the vast majority, enforcement efforts would be easily justifiable and would not be unpopular. In reality, when set at the 85th percentile, the limit was simply a surrogate for the level of apparently acceptable risk to everyday drivers, not an objective measure of the actual risk.

Currently, most authorities now apply "expert systems" to set limits that take into account a range of factors, such as the nature of roadside development, the ongoing majority of councillors that the proposed changes should continue to be supported.

The issues then landed on the desk of the state minister for roads. He did not support the proposals and refused the request even though the state road authority supported them. The reasons for this are not known, but concern about a potentially negative public reaction is likely to have been a major factor. It is notable that a minister would not approve a safety recommendation from his chief safety advisors and would not accede to the wishes of an elected council seeking to improve levels of safety on roads for which they were legally accountable.

Some 3 years passed and there was a change of government. The shire continued to make its case and to hold together its councillors in support, while running local public information campaigns in support of the proposed measures.

After the new government was in place for only 12 months, approval for the proposed speed limit changes was granted.

When the shire tested local public opinion before implementation, levels of support for the proposals exceeded 80%. After implementation, support increased to 91% for the rural road speed reductions and 81% for the residential area reductions.

What is striking here is that one person (the minister) effectively delayed these safety measures for some 3 years. The presumed concerns about public acceptability were shown to be unfounded. So often, when change is proposed, governments (and certainly individual politicians) are well behind the evolution of public opinion. This is often the lot of innovative policy in the road safety field.

mix of traffic (particularly pedestrians and cyclists), the frequency of driveways entering/leaving the road, and so on. These empirical expert systems facilitate speed limit decisions during the consultation process between agencies at different levels of government and with the community. They are a useful tool for reaching and defending decisions but are no more an appropriate match with available levels of protection than their predecessor. They are more of a surrogate for balancing apparently competing objectives in the road transport system.

9.9 SOCIAL CONTEXT OF SPEED BEHAVIOUR

9.9.1 DAILY RISK TAKING

Helen Wells is a criminologist in the United Kingdom who specialises in the *everyday crimes of the law abiding*. In an excellent recent book she has interpreted the controversy around speed enforcement via the widespread use of speed cameras within the social context of the way in which day-to-day risks are taken in a modern society.[99]

One of her most telling points is that exceeding the speed limit is endemic; it is *normal deviant behaviour.* Almost all drivers exceed the limit at some stage. A small proportion exceed the speed limit regularly, an even smaller proportion exceed it by large margins, while most exceed it by small margins on a relatively large number of occasions. Wells reviewed survey results and reported that while the majority of drivers admitted to speeding—and some 40% have received tickets for speeding— over 80% considered themselves law abiding!

Australian data from on-road measurements report that at least 80% of vehicles travelling over the speed limit do so by less than 10 km/h, with only 1% more than 20 km/h over the limit. Survey data reveal that about one-third to one-half of drivers report that low-level speeding is unacceptable, despite nearly 90% admitting doing so![209]

A recent survey of Australian drivers reported that it is "common for drivers to think that other drivers who drive faster than they do are a safety threat but they mostly see their own driving as being under control and therefore safe enough."[210]

As we have shown earlier in this chapter, speed is a risky behaviour—faster speeds (higher mean and 85th percentile speeds within measured speed distributions) result in a higher frequency of casualty crashes with higher severity levels. What is poorly understood by policy makers is that this is because the risks are aggregated across the entire road transport system—large numbers of drivers each accepting a small additional risk will, by statistical weight of numbers, lead to more crashes than if the entire road transport system operated at lower speeds.

It is an actuarial risk that an insurer would readily understand, but at the level of a given individual driver on a given individual trip, the additional risk is unlikely to lead to harm. The likelihood of a casualty crash on any individual trip for any individual driver is extremely small, and while driving at a small amount over the speed limit may double that risk, we need to acknowledge that *doubling an extremely small risk still leaves a very small risk.* So, *at the level of an individual driver on any individual trip*, the daily experience of speeding may be technically criminal, but it is not demonstrably dangerous or immoral.

Living in a modern society involves daily, almost continuous, risk taking: whether or not to smoke, whether or not to consume alcohol or illicit drugs, whether to buy or sell shares in the prevailing financial market, whether to change jobs in an adverse employment market, and so on and so on. Driving similarly involves near-continuous risk-taking decisions: whether or not to overtake, whether or not to stop and yield or to clear the intersection late in the yellow phase, and so on. Decisions involve weighing potential costs and benefits. In the case of low-level speeding there is no contest. Daily experience attests to the risk being small, while the benefits of catching the lights before they change, of passing a slow-moving truck, of a perceived time savings in a time-pressed journey are immediate and gratifying.

Many attitude surveys confirm this analysis. In Chapter 4 we presented data from the United States, Australia, and New Zealand that showed that while drivers gave socially desirable answers to questions about speeding, large proportions were prepared to admit to regularly engaging in the behaviour.

The greater risk that most drivers are concerned about in exceeding speed limits is not the possibility of a crash but the possibility of a speeding ticket. By introducing intense enforcement of speed limits, particularly automatically via speed cameras, governments have, in Helen Wells' terms, moralised "normal deviant behaviour." Thus, enforcement is demonised. As we explored in Chapter 8, the *dilemma of the commons* can be seen to be front and centre. Speed moderation by all drivers at all times requires embracing the proposition that the resultant benefit to society is such as to warrant forgoing the immediate personal gains that otherwise accrue! No wonder we are not succeeding.

9.9.2 BRIEF HISTORY OF SPEED ENFORCEMENT STRATEGIES AND PRACTICES

Speed limits are typically seen by drivers as advisory speeds for all but the most adverse conditions, not as upper limits that should not be exceeded. Historically, police officers have not been keen enforcers of speed limits, other than for extreme speeds. Police do not favour enforcing unpopular laws and, in their capacity as ordinary drivers, tend to share the view of everyone else that low-level speeding is not dangerous behaviour.

One prominent feature of speed limit enforcement has been, and in many parts of the world still is, that it is subject to an (unwritten) enforcement tolerance, typically of around 10%, but usually rounded up to 10 km/h for speed limits below 100 km/h. Thus, in an urban area with a limit of 60 km/h drivers would not be ticketed unless they were observed travelling at 70 km/h or more. In Australia at least, part of the rationale was that the vehicle design rule governing speedometers—made at a time when the measurement device was mechanical—required a demonstration of their accuracy within ±10%, so that in theory, a ticket for an offence below a measured 70 km/h in a 60 km/h zone could be challenged. This was also a comforting stance for officers who did not believe that low-level speeding was dangerous.

The principal enforcement strategy was to focus effort on the seriously deviant speeders, those at speeds well above the speed limit. The combination of an enforcement tolerance and a practice of focussing on high-end speeders reinforced in the minds of the driving public the view that it was OK to exceed the speed limit by a moderate amount. One of the authors of this book observed firsthand such an example while conducting training for the police in New Zealand on the implications of research evidence for effective speed enforcement. The urban limit in New Zealand is 50 km/h, but with a 10% tolerance motorists were never ticketed unless measured at over 60 km/h. A woman, ticketed at 61 km/h, wrote a letter of outrage to the daily press, complaining of the injustice of being treated as a criminal when she was only "1 km/h over the limit," despite the reality was that she was 11 km/h over the posted limit. What is on the speed limit sign is not what people consider the limit; the limit is what they have learned that they can get away with by official sanction!

Silcock and his colleagues in the UK conducted one of the most extensive surveys of drivers, including home interviews, focus groups, and in-car video data collection.[211] They found evidence of an internalised speed limit; drivers reported choosing their speed depending upon the prevailing circumstances of weather, traffic density, and road design rather than the posted limit. Posted limits were frequently seen as being inconsistent and generally lacking credibility, and almost all drivers understood the prevailing police enforcement tolerance.

Gradually, as the evidence accumulated relating risk to the entire speed distribution, albeit an exponential relationship, governments began to see the wisdom of a risk-based enforcement strategy. A large number of people each taking a small additional risk results in more crashes in total than a small number of people each taking a large additional risk. Some governments, for example, the government in Victoria, Australia, removed the enforcement tolerance, massively increased enforcement intensity, and added cameras to the enforcement toolbox.[174,212] The wisdom of the risk-based strategy was evident in the reductions in casualties that followed,[200] but anything but evident in the controversy the change in strategy generated. David Marr described it thus: "The mob is restless and the shock jocks are howling."[4]

Helen Wells wrote that whether or not the speed camera program in the United Kingdom reduced road crash casualties (and it did) was almost irrelevant and certainly subservient to its role in polarising the community. She argued that much damage was done to community trust in scientific evidence and to the credibility and standing of scientists as the "war" raged. The National Safety Camera Program enabled local authorities to work with police to install (fixed) speed cameras with the revenue from tickets initially hypothecated to cover the costs of the enforcement, with any surplus going into further enforcement. By 2007 the hypothecation ceased and the fine revenue flowed into consolidated revenue, a move that dramatically reinforced the widespread belief that speed enforcement is more about revenue raising than road crash injury reduction.[99]

The evidence of effectiveness in casualty reduction continued to be disputed by those vehemently opposed to the enforcement of other than extreme speeding. Groups such the Association of British Motorists recruited like-minded academics to produce "evidence" to muddy the waters, while others such as Motorists Against Detection (MAD) acted as vigilantes to vandalise cameras. Funding for the camera program from the national government completely ceased in 2010. In doing so, the responsible government minister said: "Another example of this government delivering on its pledge to end the war on the motorist."[99] In all likelihood it was more a statement of political appeasement to a vocal body of electors!

The endemic nature of low-level speeding coupled with the intense unpopularity of risk-based speed enforcement places governments in a very difficult position. Implementing what the evidence confirms is effective in reducing road crash casualties requires a high level of political courage, far more than many governments can muster. Retreat from the widespread use of speed cameras as an enforcement tool is not confined to the United Kingdom. Cameras were removed from British Columbia, Canada, in 2006. By late 2009 the citizens of 11 towns in the United States had voted to have speed cameras removed, and in April of that year a technician servicing a camera in Arizona was allegedly shot dead![213] Recent survey data from Minnesota in the United States show substantial support for speed cameras, but only in school and roadwork zones.[214]

Nor has Australia been immune. When in opposition, one of the major political parties in Victoria played to the vocal opposition to speed cameras and promised to remove cameras and restore enforcement tolerances should they be elected.[215] When they assumed government some years later, they requested an independent audit of Victoria's speed camera program, performed by the Victorian Auditor-General's Office, a neat sidestep given that evidence of the safety effectiveness of cameras had continued to build. The audit report confirmed that, considered as a whole, the program was reducing casualties and should be maintained.[216] In a clever political move, the government announced that it would publicise all camera locations, including mobile cameras, on a weekly basis and would appoint a special office to investigate complaints about unfair camera enforcement. As time passed, and as general financial pressure on the government increased, the number of hours of speed camera enforcement dramatically increased and the fines applied also increased. The speed camera revenue is not hypothecated to traffic safety but accrues to general government revenue for expenditure as the government sees fit, again reinforcing the revenue-raising belief. A very similar story can be told for the neighbouring state of New South Wales, with the major difference being the latter's hypothecation of camera revenues.

These examples are not intended in any way to denigrate government decisions—past or present, in Australia or elsewhere. Their sole purpose is to illustrate how intricately connected are the endemic nature of speed behaviour, the unpopularity of the enforcement of low-level speeding, the prevailing road use culture, and the interconnected impacts of these elements upon the complexity of policy and decision making. However, as Josh Gordon concluded his article: "The moral is that something as serious as road safety should be placed above politics, however much motorists detest speed cameras."[215]

9.9.3 WHAT ENCOURAGES US TO SPEED?

We have already touched upon two of the important factors that encourage speeding (at least at low levels):

- Instant rewards almost inevitably follow individual speeding events, such as passing a truck that was impeding us, clearing an intersection before the lights changed to red or a potentially conflicting vehicle arrived, saving precious seconds on a time-pressed journey, the pleasure of a thrill, etc.
- A lengthy history of a 10% enforcement tolerance taught us that the limit was not a true limit, which must mean that the government cannot really regard speeding at low levels as dangerous.

As David Shinar argues, speed is *logically related to mobility*—go faster and get there sooner—and *subjectively related to pleasure*.[13] Further, there are pervasive social factors that underpin speed behaviour.

The first relates to vehicle design. Improvements in vehicle engineering have almost continuously increased engine power, acceleration performance, and top speed capability of all makes and models while at the same time reducing the feeling of speed through an absence of noise and vibration. No serious design attempt has been made to offset the impact of these pervasive increases, presumably because performance is a fundamental sales feature. While there is a veritable raft of vehicle design rules for safety, not a single rule relates to power, acceleration performance, or top speed capability. It seems that speed moderation through vehicle design is off-limits to both the industry and governments in the world of safety regulation!

Even more tellingly, the speedometer, the primary tool provided to the driver to monitor travel speed, is calibrated to show speeds that are illegal on all but a very few of the world's roads—top speeds shown on the instrument range from around 220 to 300 km/h, at least double what is permitted on most high-speed roads (Figure 9.4). Moreover, the range of legal speeds—between 30 and 110 km/h, depending upon the road section—occupies half or less of the available instrument space, making the driver's task in judging speed relative to the posted limits unnecessarily difficult. It is not uncommon to hear motorists ticketed for low-level speeding complain that they were caught for an offence that was "barely the thickness of the speedometer needle" above the limit. If the primary purpose of the speedometer is to enable the driver to comply with posted limits, it is clearly not fit for purpose!

Why is such an inappropriate design tolerated? At one stage in the United States a notice of intent of vehicle design rule making was published through which the top speed shown on the speedometer was to be limited to 80 mph (approximately 130 km/h). The proposed rule never came to enactment, being successfully challenged on the grounds that there was no evidence to establish that a safety benefit would accrue. Unfortunately, this left the fitness for purpose question unaddressed. Despite initial support by one Australian state, an attempt by one of the authors to obtain national funding for a simulator-based study of the effects of different designs of speedometer on speed choice was rejected on the grounds that such a study was quite inappropriate!

FIGURE 9.4 Speedometer design. Note that despite a maximum speed limit of 110 km/h in Australia, the speedometer goes up to 260 km/h.

The more fundamental question is what does speedometer design tell us about the way cars are marketed and sold? In Section 4.2, on car culture, we explored the masculinity, aggressiveness, and power connotations contained in the naming and representation of cars, and there is no doubt that vehicle manufacturers compete for the power and speed market. For example, a press advertisement for a sports model of a leading car manufacturer claimed that the car could "DEFY PHYSICS" and urged drivers to: "Hammer the straight. Scream through the corner". Such hyperbole is hardly supportive of a safe road use culture! This type of advertising remains an everyday event in television advertising and in print media. Complaints from concerned citizens and from government agencies have failed to have an impact. The New South Wales government's roads agency lodged over 40 unsuccessful complaints about television advertisements for vehicles that, in its view, promoted inappropriate speed behaviour.[209]

The automobile industry in Australia hides behind an industry-derived, voluntary code of advertising practice that defends most complaints of the kind cited above with the facile statement that the advertisements portray off-road events that are clearly fanciful and could not, therefore, induce inappropriate behaviour! The "fantasy" defence has been called the industry's "get out of jail free card," which illustrates the poor regard in which the code is held by traffic safety professionals and concerned citizens.

The code, introduced in 2002, requires that "advertisements do not depict, encourage or condone dangerous, illegal or reckless driving" and, in particular, "are not to promote a vehicle's speed or acceleration capabilities" whether explicitly or implicitly.[217] A formal evaluation of the impact of the code has been undertaken. Nearly 100 advertisements after the code was introduced were compared with a similar number before. A decrease in the frequency of performance and fun themes was found, but no change in the frequency of themes relating to speed, power, and acceleration.[218] In a second study three advertisements that had been the subject of complaints (two of which had been dismissed) were shown to a large sample of drivers. All three were found to convey messages contrary to the code.[217]

Most disturbing, though, is the process of dealing with a complaint. The complaint goes to the Advertising Standards Codes Board, which invites the advertiser and its client to respond to the complaint before making a ruling. Even if the ruling is that the advertisement encourages unsafe behaviour, the board can only recommend its withdrawal. If the advertiser declines to comply, nothing further happens. In any event, the time the complaints process takes ensures the advertisement has run the course of its normal effective life before a ruling emerges. Not only do complaints rarely succeed, but when they do, success is purely symbolic!

The inadequacy of industry self-regulation, particularly around marketing and advertising, goes far wider than the car advertising. An Australian television program using advertising industry experts to examine advertising techniques explored an advertisement for a universally known soft drink that used a well-known celebrity to claim that the product could be enjoyed by children with no adverse impact on their dental health. Complaints about the advertisement were rejected by the Advertising Standards Codes Board on the grounds that the fine print included a statement that this claim assumed teeth were cleaned after every drink of the product, an indefensible caveat!

It is instructive to draw a parallel between the automobile industry's denial of the crash risks associated with speeding and the tobacco industry's (initial) denial of the health risks associated with smoking. Moral pressure on the tobacco industry had little effect and extensive government action was required. Even now, the tobacco industry's primary defence is that it is an individual's right to smoke and his or her personal responsibility if he or she chooses to do so. The automobile industry holds an apparently identical position with regard to the provision, and active promotion, of vehicles designed to exceed safe speeds by huge margins: it is the individual's responsibility to use the power, acceleration, and top-speed capability appropriately. Moral pressure is having little, if any, effect on the industry.

Lewis Hamilton, the champion Formula One racing car driver, was apprehended doing burnouts in a sponsor-provided car on a public road in Melbourne the night after the Formula One race had been held in 2010. The public concern was scoffed at by fellow champion racing driver Mark Webber (an Australian), who accused Victoria of being at the forefront of "nanny states."[219]

Before we are accused of being antiautomobile and of taking cheap shots at the industry, let us answer the chicken-and-egg question of whether the industry is responding to the demands of its market or whether it created that market. As Moeckli and Lee state: "A car's design is as much a response to a driver's fantasies of power, control and speed as it is to the utilitarian components of travel."[220] In Chapter 4 we explored the culture of the car and of driving. While the automobile industry has resisted all efforts to curtail its response to the market, the simple fact that governments have never pressed for safety design rules limiting speed capacity is the most cogent evidence that governments are responding to what they perceive the community wants at least as much as reacting to industry pressure.

This is where the parallel with the tobacco industry breaks down, at least to a degree. While lots of people smoke and lots of people speed, the community support for restrictions on smoking in public places, the carriage of health warnings on cigarette packets, and similar measures is strong, while the support for strong measures to achieve speed moderation is tepid.

9.10 HOW MIGHT WE ACHIEVE WIDESPREAD SPEED MODERATION?

We have devoted an entire chapter to exploring speed behaviour because speed moderation is—until we design and implement a truly safe road and traffic system—the most effective measure that we have in our armoury of traffic safety countermeasures, and it is the least well understood. It seems hard to avoid the conclusion that most governments appear to have reached—that the barriers to achieving system-wide speed moderation are insurmountable. We argue that such a conclusion is wrong because the traditional approach is fundamentally flawed.

Our traditional approach has comprised the following:

- The introduction of a wide range of rapidly changing speed limit zones (around schools, strip-shopping areas, approaches to rural towns, and so on) as a compromise between the desire for higher limits for maximum mobility and the need to address the safety threats on specific sections of road—a practice that increases confusion as to what zone a driver is in at any given point in his journey.
- Public education to inform the community that speeding, even at low levels, is risky—information that is discordant with everyday experience where low-level speeding is endemic and no harm occurs (at an individual level, most of the time).
- Risk-based enforcement, using both police officer surveillance and automated camera technologies to achieve intense enforcement in the belief that behaviour change via specific deterrence would achieve knowledge and attitude change—a strategy that flies in the face of popular belief and is intensely unpopular.
- Reducing enforcement tolerances as part of the risk-based strategy—a policy that further increases resentment.
- Directing fine revenue into consolidated revenue rather than it being hypothecated to traffic safety measures—reinforcing the belief that the enforcement is not only unjustified but is a government means of raising revenue.

What we have in the above list is a disjointed series of tactics based on false assumptions about the fundamental reasons for the speed distributions we observe. What we need is a comprehensive, integrated strategy to ensure speed distributions are appropriately matched to the levels of protection provided within our road transport system. The strategy must be built upon a deep understanding of speed behaviour and must be designed to engender community support and commence a long-term culture change process.

We must address every aspect of the environment in which speeding occurs and overcome each barrier to change through an integrated package of changes. The following initiatives have to be taken.

9.10.1 Road Design and Appearance

We need systematically to reexamine current design principles and practice. We might follow the lead of the Dutch, who have a long-term goal of achieving "self-explaining" roads comprising only three categories—streets where people live, roads that link areas where people live with shopping and service centres, and highways that link cities and towns. By ensuring distinctive visual appearances for each category (hence the term *self-explaining*), the Dutch believe behaviour will eventually match the road function because people will know from their appearance what is expected.[221] Residential streets, for example, may well be replete with trees and barriers that, far from being a safety hazard, serve to discourage speed. In New Zealand a successful attempt has been made to create such an environment, an experiment that confirms that the concept is applicable across cultures.[222] In Australia, at present, we rely on a range of engineering measures, such as road humps, mini-roundabouts, slow points, and the like, to reduce speeds on local, residential roads where inappropriate initial design has created an inappropriate speed environment.

Of course, we cannot remake the road environment overnight, but we can promulgate planning and design guidelines that will move us toward the ideal over the long term. While we apply a road hierarchy nomenclature at present, it is based more on current function than design fundamentals, and we observe a range of variations of design that make it highly doubtful that the community can distinguish the classes of road or what is expected of them with any accuracy.

It is simply not acceptable to say that the road system exists and cannot be changed. That is a recipe for continuing current practice in perpetuity. We must set about developing new design guidelines and a plan for their gradual implementation.

9.10.2 SPEED LIMIT SETTING AND SPEED LIMIT SIGNING

We need to match the limits to the levels of protection offered. This is something we can do right now. Since we already have a credibility problem, we will almost certainly have to stage the reduction of limits on roads where minimal protection exists. A risk analysis approach should be used to identify the roads where reduced limits will give the greatest short-term results. Public education will be needed to explain the rationale for the changes, and enforcement will need to be carefully managed to avoid an adverse reaction in the settling down period. What we must stop doing in our desire for high-speed mobility is continue to permit inappropriate speed limits to exist. High-speed mobility requires designed high-speed roads.

A related approach is to vary the speed limit according to the traffic volume or weather conditions. We know that drivers slow down in adverse conditions, so the credibility of this approach is likely to be high. The use of variable message signs (VMSs) can become widespread and may be a more acceptable start to the process of matching all limits to the prevailing levels of protection.

We need also to ensure that all limits are adequately signed. One of the major criticisms of intense enforcement of low-level speeding is that the driver was unaware of the prevailing limit as "they change rapidly from section to section." It is a simple matter to ensure all speed limits are signed using flashing LED displays rather than static signs. This practice occurs spasmodically at present in cities like Melbourne, particularly in school zones on major roads. The standard agency objection of cost, given funding constraints, has no merit when governments continue to amass substantial revenue from speeding fines. Governments have a duty of care to ensure that every reasonable step is taken to advise motorists of the prevailing limit. Enforcement must not only be fair, but also be seen to be fair.

A longer-term option is the installation of intelligent speed assist (ISA) devices in all vehicles. These devices warn the driver when the speed limit is being exceeded by combining GPS locators with a database of speed limit signs. The technology exists and, if it can be retrofitted, may be a more cost-effective option than replacing all speed limit signs with active LED signs.[223] With technology to register a vehicle's speed continuously, one might even consider an approach analogous to London's congestion charging scheme. Individual vehicles could receive a regular account where they are charged for their time above the speed limit, on an increasing scale if the behaviour is repeated.

The critical point to note is that neither option is, to our knowledge, under investigation. Governments are not delivering an adequate duty of care to underpin intense speed limit compliance efforts.

The Dutch also focus on setting the minimum number of limits possible— describing their limits as SACRED (*sa*fe and *cred*ible). Ensuring credibility is a vital feature of a system that is seeking to achieve community acceptance. Speed limits must be seen to be appropriate and fair. Self-explaining roads and SACRED speed limits clearly go hand in hand. Credibility, of course, is a double-edged sword. Where the built level of protection does not support a high speed but the appearance of the road environment suggests it should, a conflict between safe and credible exists. The Dutch philosophy is to remove the conflicts by matching appearance to level of protection.

9.10.3   VEHICLE DESIGN AND PROMOTION

There is little value in decrying the use of speed and power in vehicle advertising without trying to address the root causes.

Governments and industry must commence a dialogue about the future of speed issues in vehicle design. As already pointed out, there is not a single safety design rule related to top-speed capability, acceleration capability, or power, despite incontrovertible evidence that kinetic energy is a function of (half the) mass times (the square of the) speed. This will not be a comfortable dialogue, at least in its early days. But start it must. The most obvious starting point is around instrument design, especially the speedometer. It would be a very simple matter to make the instrument better fit its designed purpose of assisting drivers to maintain compliance with speed limits. The installation of ISA (described above) would be another simple step for manufacturers to assist drivers to comply with speed limits.

Influence over vehicle advertising standards is a secondary avenue along the path toward more appropriate vehicle design. A dialogue around advertising could take place simultaneously with the dialogue around design standards. The threat of mandatory advertising standards to replace the demonstrably ineffective voluntary codes that currently exist may be a distasteful but necessary catalyst to bringing the parties to the table.

9.10.4   RISK-BASED ENFORCEMENT

The evidence that risk-based enforcement is the most effective enforcement strategy is extremely strong. However, such enforcement is also extremely unpopular. If the road system, the speed limit system (and its signing), and the vehicle design issues were all resolved, much of the controversy would, in our view, dissipate. But these fundamental issues are not going to be resolved quickly. In the interim, risk-based enforcement must continue, but under a set of strictly applied conditions designed to minimise community concern:

- The rationale for all enforcement locations should be publicly explained and transparently linked to crash or crash risk data.
- The punishment should fit the crime. Low levels of speeding should be treated with low-level fines. Perhaps the first offence should result in a suspended fine and a warning letter, with a second low-level offence attracting the suspended fine from the first offence plus a fine for the second offence, plus the allocation of demerit points. We are not proposing this as the ideal system, but simply pointing out that we need to convince the motoring public that our interest is not in revenue but in speed moderation for safety.
- All fine revenue must be publicly and transparently allocated to traffic safety measures with regular explanations of the allocations and their outcomes.
- An independent auditor should be appointed to act as a community representative to hear complaints—a sort of speed ombudsman.

9.10.5 PUBLIC EDUCATION

Public education is a primary tool in current speed moderation strategies in Australia, but needs supplementing with additional messages. One common theme is to raise the level of knowledge about the effects of small increases in impact speed on crash likelihood and injury severity. A second theme is to emphasise the probability of detection and punishment. The third common theme is to stress the effects of severe injury on lifestyle. While all are appropriate, they run the risk of reinforcing the unpopularity of intense enforcement of low-level speeding.

Additional educational themes should include explanations of the mismatch that currently exists between speed limits and levels of protection, statements of intent to improve road design and vehicle design over time, public scrutiny of vehicle advertising, and promotion of a new approach to managing offences, especially first low-level offences, with commitment to the hypothecation of all fine revenues to traffic safety purposes. The impact of education will be enhanced if all speed limit signs are active, that is, highly visible LED signs.

A broader program to provide social approval for those who comply with limits and to socially isolate persistent speeders is worth consideration but, without other changes to current strategies, may lack credibility. The city of Joondalup in Western Australia ran a campaign where people who pledged to keep to speed limits received a "Safe Speed Promise" sticker to display on their cars. Unfortunately, only 1,000 people out of a population of 200,000 signed up, and the majority of these were over 50 years of age.[224]

9.11 CONCLUSION

With this more complete understanding of the place of speed in our culture and the interplay between speed behaviour and road and vehicle design, we can quickly see how inadequate our sole current strategy—of continuing to rely on intense enforcement supported by public education—is and why it raises the ire of the motoring public. We must not retreat from intense enforcement, but we must effectively address the issues that make it unpopular, and these involve addressing all the system weaknesses that we have described.

We must develop a truly integrated strategy that includes road design, vehicle design, speed limit setting, and intense enforcement that is seen as fair and reasonable. Yes, the barriers to system-wide speed moderation are considerable, but if we can achieve it, and we believe that we can, we will be well on our way toward the ideal of achieving zero deaths and disabling injury within our road transport system.

10 Confronting Complacency

10.1 WHY TRAFFIC SAFETY LACKS BOTH A COHERENT CONSTITUENCY AND COMMITTED LEADERSHIP BY GOVERNMENT

In Chapter 1, we stated: "It is impossible to escape the conclusion that we have the level of trauma that we, as motorised societies, are comfortable with" and that we were going to try to "understand why complacency rules."

Having presented our evidence, it is now time to summarise our understanding. The following list does not imply a temporal sequence of events, although each element in the list is intimately connected to the others. The list highlights the complexity of the total social and cultural environment in which transformational change is necessary if traffic safety efforts are to meaningfully set a target of zero disabling injuries and deaths.

1. Modern Western societies are highly dependent upon an efficient road and road transport system to achieve their overarching goals of continuous economic growth and (largely) unfettered personal mobility—two of the keys to the way these societies perceive quality of life and standard of living.
2. Moreover, pressure on infrastructure is building. The freight task is growing faster than the economy (at least in Australia). Truck traffic is growing at twice the rate of car traffic, and the trucks are becoming larger and more powerful.
3. Historically, we designed, and operated, a fundamentally unsafe road transport system because, as motorised transport was a revolutionary development, the necessary fundamental knowledge to do otherwise simply did not preexist. We evolved inappropriate policies, practices, and designs from an unmotorised era of personal transport because we had nothing else to go on and struggled to react to the rate of growth. While we have made substantial improvements over a range of specific elements, the system remains inherently unsafe, exacerbated by the increasing diversity in vehicle mix and mass.
4. Cars and driving are deeply embedded in modern, Western culture, helping to satisfy such personal goals as independence, individualism, freedom, status, pleasure, and many more besides. Frequently, television and film sequences—and advertising pitches for a wide range of products—portray the positive, joyful and exciting aspects of motoring.
5. As individuals, our daily road use is overwhelmingly accomplished without adverse events; *it is indeed a low-risk activity at the individual trip level.* We have come to believe, from this personal experience, that we are individually in control of our own safety, that crashes are predominantly the fault

of those involved, and that a quantum of road trauma is an unfortunate but necessary price society (that is, others) must pay for our personal mobility.

6. The high number of disabling injuries and deaths is *a direct result of the staggeringly high volume of road use in a fundamentally unsafe system*. As long as this combination exists, no other outcome is possible, despite each trip involving a statistically low risk at the individual trip level. As Allsop pointed out, the risk of a casualty per hour of driving, measured at the aggregate community level, is seven times greater than that for any other widespread daily activity.[11]

7. Given that behaviour is critical in an open-loop system, error is commonplace. What is surprising is just how well, on average, road users cope with a system that is not error tolerant. All that we tend to hear, unfortunately, is how badly we behave!

8. Of course there is misbehaviour—alcohol and drug-affected driving, speeding, aggressive driving, and so on—and we must continue to limit bad behaviour as best we can through education, legislation, and enforcement. Technological advances, such as alcohol ignition interlocks and intelligent speed assistance systems, hold promise as a tool of the future. We now know that illegal behaviour is a causal factor in about 40% of fatal crashes, but only about 10% of casualty crashes.

9. There is widespread ignorance among the public of the true extent of road trauma; not only is the number of deaths grossly underestimated, but there is almost no appreciation of the number—or of the widespread and long-lasting impacts—of serious injuries.

10. Most governments now understand and, at least at an intellectual level, accept *Safe System* concepts and principles that involve designing and operating an error-tolerant system while containing illegal behaviour. However, implementation is faltering and hesitant at best. Unlike the case in industrial safety, governments see little or no opportunity to transfer to others the cost and accountability for system-wide change to radically improve traffic safety in the inherently unsafe system that has evolved.

11. Governments are the owners, operators, regulators, and managers of the road transport system. They regard the potential cost to themselves of comprehensive Safe System changes as prohibitive and the risk of liability for specific instances of failure as disturbingly high. Moreover, their taxpayers are not clamouring for change, and certainly not for either increased expenditure or greater restraints on personal access and behaviour. Not surprisingly, governments—and the industries directly dependent on cars and driving—overstate their safety achievements and do little or nothing to tackle community ignorance or counter the prevailing belief in fault and blame.

12. Consequently, most governments tend to take action only at times of crisis—an apparent spike in road crash deaths—and, despite current knowledge, all too often apply an educational/regulatory/enforcement solution rather than taking a systemic approach, because it is consistent with popular belief, is of lower cost, and citizens can see immediate action. Many, if not most, Western governments remain reactive—occasionally proactive, but certainly not generative—in their safety policy setting.

13. Driving is entrenched as a right, rather than a responsibility. The almost continuous personal rewards that flow from our moment-to-moment driving behaviours far outweigh the sacrifices we might have to make if we were to drive with the health of the community as our primary objective.

14. Citizens are routinely shown only the tip of the road trauma iceberg and are told regularly of the progress their governments are making, with respect to the rest of the world. The daily media diet of crash reporting continuously reinforces the "blame the victim" mindset. Official traffic safety strategies contain targets for reductions in disabling injury and death, with no indication of the (officially acceptable) levels of tragedy that will occur during the lives of these strategies. The message put out is that we are doing well and will continue to improve. *The real message should be that the societal risks of daily road use remain in the top rank of public health problems in the twenty-first century and demand a commensurate response.*

10.2   UNDERSTANDING THE CHALLENGES

We have just presented a formidable catalogue of barriers that make a spontaneous uprising of community demand for the eventual elimination of disabling injury and death from everyday road use an almost laughable objective. How, then, might forces sufficient to generate transformational change in public policy and political action arise? Before presenting our view of the steps needed to overcome complacency, we should consider the essential prerequisites for transformational change.

Three necessary conditions must coexist before new, or changed, public policy can be formulated. Policy makers must

- *Know* that a problem exists and *accept* that it must be addressed.
- *Be able* to address it; that is, they must assess it as both politically feasible and practically implementable.
- *Want* to initiate change, which requires persuasion more than it requires knowledge or understanding.

10.2.1   KNOWING AND ACCEPTING THE PROBLEM

Knowing that there is a problem is the first step. Problems do not exist just because there is evidence; problems only exist when there is an alignment between perceptions, beliefs, and attitudes, and the facts. Perceptions, beliefs, and attitudes are subject to polarisation by vested interest groups, determined to influence the outcome of public debate.

Kotter's first principle for businesses seeking transformational change is to establish a sense of urgency, and that is equally applicable in public policy.[225] Acceptance of the urgency of a problem is at least as important as a consensus that the problem exists. Lindblom and Woodhouse insightfully wrote: "Policy making resembles a primeval soup—action occurs fitfully as problems wax and wane and pressures come and go."[175] Let us consider some examples.

In 1955 the number of road crash deaths in Australia exceeded the aggregate number of deaths from all forms of infectious disease for the first time. This statistical fact was drawn to national government attention through a question asked in federal parliament, with an accompanying plea that Australia's major government-funded research body* address the matter urgently. The prime minister of the day dismissed the plea as preposterous, and there the matter ended![226] The statistical fact had no traction; it was simply not accepted as important.

Fifteen years later, the number of road crash deaths in the state of Victoria exceeded 1,000 for the first time in the state's history. Going into four figures was like breaking a psychological barrier, and this time, the statistical fact became an issue of public disgrace. The most popular daily newspaper in Victoria started a vigorous campaign demanding action. The slogan was "Declare war on 1034." It is worth noting that the editor of this newspaper was a close friend of a leading trauma surgeon who spent much of his time operating on crash victims and who was personally committed to promoting preventive action.[227] The campaign helped build a climate of concern that contributed, in turn, to the passage of the world's first mandatory seat belt wearing law, a law with a dramatic immediate effect and a global legacy.[78]

The difference between the 1955 fact presented blandly to federal parliament and the fact of the 1970 peak in deaths in Victoria was that the latter gained traction through vigorous public discussion. The public discussion occurred because of the committed leadership of surgeons and their link with a media executive. The lesson to be drawn is the power of influential champions in raising issues publicly. While knowledge matters, it is a necessary, but not a sufficient, condition for public policy action.

In 2012 the Australasian College of Road Safety (ACRS) and the nongovernment coalition 33,900 jointly put a case to the national Productivity Commission for an inquiry into the management of traffic safety in Australia. Their case was based on presenting knowledge about the impacts of road crashes predominantly upon those of workforce age and presenting findings that road crashes comprised the greatest safety risk to companies in Australia. The approach was rejected, again illustrating that facts alone have little traction.[228] The ACRS and 33,900 are not yet effective champions in stimulating public debate; they, as yet, have little traction with the community. The lesson is that traffic safety advocates need to partner with experts in social marketing if issues are to become matters of public debate.

As we have already seen in Figure 8.2, the four major traffic safety policy changes in Victoria between 1970 and 2001 followed an apparent spike in deaths, each painted by the media as a crisis. Each spike was relative to the total to the same point in the prior year. Unfortunately, in a crisis that is of short duration, the public demand is for immediate action, not for long-term systemic changes. The lesson for traffic safety professionals, however, is to have available one or more measures of known effectiveness but previous unacceptability to be promoted to government when a crisis—real or apparent—provides the opportunity. A media-reported crisis is an excellent indicator of a problem on the cusp of acceptance.

---

* The Commonwealth Scientific and Industrial Research Organisation (CSIRO).

We need also to understand that evidence is not the only currency exchanged in a public debate. Hinchcliff analysed newspaper reports and television news coverage of young driver safety and the prospects of nighttime and passenger restrictions being introduced to counter these high-risk factors.[229] He examined media coverage over a 3.5-year period to the middle of 2004, when the policy options were being debated, and found over 250 newspaper items nationally and almost 40 television news items in Sydney alone. There was widespread support in the media coverage for the moral imperative—the problem was reported as severe and governments were being urged to do something about it! The problem appeared accepted. However, the scientific evidence of risk and arguments about the potential impacts of the proposed restrictions (the solutions) on young drivers' mobility got almost equal media coverage. It seems the solutions being proposed were not accepted, although the problem was. The lesson is that we must not just present scientific evidence to support proposed solutions. We must examine potential barriers to implementation and develop a strategy to overcome these if we are to gain true acceptance.

In Chapter 6 we lauded the evolution of Safe System as the conceptual approach underpinning the development of modern traffic safety strategies. We also noted a persistent gap between knowledge of Safe System principles at the political and bureaucratic level and the acceptance of their implications for policy and practice. We speculated that the apparently high costs of implementation and the need to accept public accountability for outcomes make road trauma "an inconvenient truth" for governments throughout much of the Western motorised world.

Myriad examples exist, but we shall cite just one here. Others have appeared earlier in Section 5.1. In May 2011, while launching the Australian component of the UN's Decade of Action, the federal minister responsible for national traffic safety efforts in Australia first congratulated authorities for adopting the Safe System approach in the current national road safety strategy, but went on to emphasise throughout the bulk of her address that the major challenges to be overcome were the "big four" behavioural issues of speeding, drugged driving, drunk driving, and fatigue![230] This immediately undermined the Safe System concepts by refocussing attention on blameworthy behaviours, and thus reinforced popular belief.

Of course, the minister did not write her own speech, just as the Victorian politicians did not write their own press releases (see Chapter 5). Science may have written the modern traffic safety "hymnbook," but as yet, political leaders are not singing from it! The community is not demanding that the politicians accept the urgency of the problem.

10.2.2 Being Able To—Political and Practical Feasibility

Knowledge is important, but accepting the need for action and embracing that acceptance are crucial to transformational change. We have to accept that in a democracy compromise rules.

The editor of the *Sydney Morning Herald* wrote in an editorial of August 12, 2012: "Serious debate … is being usurped by a managerial approach to politics. Political parties obsessed with polling and focus groups, and politics played as a blood sport, with its attendant tribalism and biffo, are undermining the quality of public discourse."[231]

While he was not talking about traffic safety, the statement is apt. In Chapter 9 we cited controversies around automated speed enforcement using cameras, controversies created by opportunistic posturing by opposition parties in response to noisy minorities. We gave examples from two Australian states and from the United Kingdom, with, in the Australian examples, radical reversions in position once the opposition parties assumed government. Scientific knowledge about the effectiveness of the intense enforcement of speeding violations was totally overshadowed by "noise" in the public and political debate.

Success in innovative policy making can be seen when traffic safety is separated from political partisanship. The Swedes made Vision Zero an all-party commitment within the Swedish parliament in 1997. In Australia, the government of Western Australia went to considerable lengths to engage politicians from across the political spectrum before submitting its new traffic safety strategy to the parliament. Sadly, there are few examples of this process of de-politicising proposed interventions.

No one can create a social movement for transformational change just by calling for it to happen and presenting scientific knowledge as the rationale, no matter how compelling. For change to be politically feasible, there must be a reasonably widespread and reasonably intense level of discontent with the status quo within society, at least among influential opinion leaders. There is no significant discontent in Australia with regard to road trauma; such discontent, as has been seen, has been limited to short periods of time around perceived crises.

In the United States, a grassroots community group was created in 1980 by a mother outraged that the drunk driver who had killed her daughter was dealt with leniently. In their first 4 years of activism, Mothers Against Drunk Driving (MADD) saw the creation of a Presidential Commission on Drunk Driving, the creation of a pool of federal funds for state initiatives to address drunk driving, and the passage of a federal law raising the minimum legal drinking age to 21 years.[232] Specific purpose lobby groups can be very effective in bringing about change if they capture sufficient support to be noticed. MADD was effective in garnering media attention and, by focussing on the tragedies drunk drivers cause, clearly held the high moral ground.

There are few Australian examples. The role of the Royal Australasian College of Surgeons in influencing the introduction of mandatory seat belt wearing is one. The pivotal influence of committed individuals can also be cited. John Birrell, the surgeon for Victoria Police in the 1960s and 1970s, insisted on taking senior government politicians with him as he attended crash scenes on Saturday nights to demonstrate firsthand the role of alcohol in serious crashes, at a time when the statistics were simply not accepted. In John's autobiography he relays how his attempts to influence traffic safety policy almost cost him his job before his confrontational approach of taking politicians to crashes began to succeed.[106] This book is recommended reading for anyone trying to understand how to influence change in an inertia-filled system.

At present in Australia, lobbying efforts are less specifically targeted. The Australasian College of Road Safety was established primarily to provide a network for traffic safety professionals and practitioners to share knowledge and experiences.[233] While it puts positions to government on issues as they arise, it rarely actively lobbies on issues on its own initiative. The more recent creation of 33,900— the Australian Road Safety Collaboration has sought to bring together nongovernment groups to press for concerted Australian action as part of the United Nations Decade of Action. Whether it will develop into an effective lobby is yet to be tested. Both groups are gentlemanly and have not studied the lessons to be learned from the success of groups like MADD or, in other fields, groups like the National Riflemen's Association in the United States.

To generate political traction, we almost certainly need to take action on a range of fronts, each with a specific target, rather than attempt to enlist community support for an all-out attack on the occurrence of road trauma. Such attempts will include the essential facts about each problem and the evidence of effectiveness for the proposed solutions, but the focus must be on creating outrage for the issue failing to be addressed and must include a strategy for overcoming any obstacles to implementation. For example, we might

- Enlist support for action on specific roads or road sections with high crash numbers. The 33,900 group has made a promising start along these lines with a push for action on a section of road in far north Queensland where local support and industry support are both high. The action plans will be anchored in the Safe System approach and will not mirror the traditional engineering black spot treatment protocol.
- Enlist support for changes to vehicle design (starting with speedometer design) and inappropriate vehicle advertising to start the dialogue needed around the role of vehicle design and marketing in speed moderation.
- Mobilise support for further systematic approaches to reducing the crash rates among young novice drivers; many parents and parent groups are already strong supporters.
- Campaign for greater action to install roadside protection systems such as wire-rope barriers by seeking media attention to focus on tragedies that result from "innocent" run-off-road crashes.

In short, by specifically listing all the evidence-based measures that currently lack community or political support, we could generate a series of targets for effective lobbying. By identifying the natural allies and the natural enemies for each measure and preparing appropriate strategies, we could succeed in generating vibrant public debates and, ultimately, the effective implementation of the key measures that would substantially reduce disabling injury and death. By seeking bipartisan support for each specific intervention before engaging in public debate, we should greatly increase our chance of success.

It is not enough to have created a body of knowledge and developed a first-rate strategy if we don't know how to get it implemented. At present, we focus our implementation efforts on government agency senior staff, senior policy makers, and the like, and we tend to base our advice largely on presenting the scientific evidence. We need also to advocate to the community and the opinion leaders, and we need to address all the blockers to implementation as well as the evidence in favour. For example, if a nighttime curfew on novice drivers is going to be effective in reducing casualties, we have to present a plan for managing the possible consequences of lack of job access, etc. We need not only to promote the safety measure but also to market a practical implementation plan.

As the marketers point out, product positioning and agenda setting are best achieved by associating the product with desirable outcomes. Product marketers would not associate speeding at low levels with a fear of punishment or with an image of a bad driver, but would seek positive reinforcements, such as, perhaps, parents role modelling for new, young drivers, mates looking out for mates, and so on.

Similarly, it is time we stopped regarding the current marketers of vehicles as the enemy. They believe they are responding to their market when they promote speed, acceleration capability, and power, and they may well be. Who cares whether there is indeed a chicken and an egg? We should assist them to explore alternative appeals that have sales pull and even offer to partially fund marketing trials. We need to embark on a journey together, not to engage in a war.

Marketers tasked with changing culture around speed would, almost certainly, avoid describing the behaviour as speeding. Labelling is nontrivial. For example, in the United States the National Agricultural Chemicals Association changed its name to the American Crop Protection Association to avoid the bad connotation around widespread chemical use; groups in favour of abortion refer to themselves as pro-choice, which appeals to individual rights, not as pro-abortion. While discussions of naming and positioning can provoke debates around ethics in marketing, we should not be coy about seeking the most effective ways to garner community support for the measures we know to be evidence based and which have the potential to reduce substantially the unacceptable burden of disabling injury and death.

To be able to form and implement new policy, governments need to see both a high level of community support and an acceptable path to implementation. It is time traffic safety professionals became interested in developing and promoting implementation plans.

Traffic safety professionals and practitioners in Australia are hampered in efforts to form alliances with others committed to the same cause, the eventual elimination of disabling injury and death. The constraints seem to have their roots in the institutional setting of traffic safety: the vast bulk of professionals are employed within government transport agencies where their actions must align with the objectives of the agency. They simply cannot join forces with victims groups or community groups to help create community outrage over either the overall level of trauma or the absence of action on a specific issue such as the protection of unsafe roadsides by the installation of barriers. They can promote Safe System concepts, but they cannot publicly discuss their lack of widespread implementation within their own agencies or by the governments they serve. They cannot rock the boat. They are limited to promoting change internally and incrementally. Similarly, most traffic safety research groups in Australia are dependent upon government grants and contracts for their research programs and must be circumspect in their public comments.

10.2.3 WANTING TO CHANGE

Knowing and accepting that an urgent problem exists is the first step in public policy reformation; political and practical feasibility of the proposed solution is the second. The third, and arguably the most critical, step is persuading political leaders to create the necessary change.

The general public is largely ignorant of the level of trauma that results from daily road use. While bureaucratic and political leaders are not ignorant of the statistics, they appear not to accept that the problem is among the few most serious public health problems modern motorised societies face. Traffic safety strategies remain incremental. There is simply no sense of urgency.

Applying current scientific knowledge of how to create a safe road transport system requires government agencies and their political leaders to accept direct accountability for safety outcomes, which in turn requires comprehensive changes to the design and operating standards and practices that have operated since the advent of motorised transport. The potential liability issues and the perceived cost of such transformational change render such change politically unfeasible. The Safe System is honoured in name only. No attempt appears to have been made to plan its staged introduction. The reader is referred back to Chapter 6 and urged to consider again the position adopted in Sweden as expressed by Claes Tingvall.

We accept that the elimination of disabling injury and death from daily road use cannot be achieved in the short term. We accept that setting targets for annual reductions in the absolute numbers of both injuries and deaths is critical in the formulation of safety strategies and action plans. What we do not accept, however, is the typical choice of soft targets based on political compromise, rather than on scientific knowledge of what is achievable right now. We regret the lack of acceptance of road trauma as an urgent public health problem. And what we strongly object to is the failure to accept public accountability for safety outcomes and the consequential unwillingness to vigorously implement Safe System principles.

Given the level of public ignorance and the absence of political acceptance of the case for transformational change, it is clear that most governments have a heightened awareness but simply have no appetite for radical change. What steps might be taken to create a political appetite? Chapter 10 addresses what can be done right now to accelerate the implementation of Safe System measures and considers the transformational change required for longer-term success.

10.2.4 PUTTING SAFE SYSTEM INTO PRACTICE RIGHT NOW

There can be no doubt that traffic safety professionals have come to accept the key principles underlying the Safe System concept: humans make errors frequently and the body has limited tolerance to kinetic energy. As a result, vehicles, roads, and traffic systems must be designed in ways that both manage kinetic energy potentials and are error tolerant. Sadly, the public has not accepted these principles; the myth of misbehaviour and lack of skill as the principal cause of road trauma is fed continuously by media reporting and vested interest groups. Illegal behaviour is a causal factor in a sizeable minority of fatal crashes, but in only a tiny minority of nonfatal casualty crashes—in total in perhaps 10% of all serious crashes (see Chapter 6).

Our road and traffic system was designed and built long before Safe System principles evolved. The road system has a long life, at least 30 years from initial construction; much of our infrastructure was built using the design principles of yesteryear. Design, construction, and operation focussed on maximum mobility at lowest cost at a time when motorisation was rampant, infrastructure funds were scarce, and safety was considered to be a matter under individual road user control. Transformational change is required to convert an inherently unsafe system into a truly safe system. Transformational change will not occur overnight; indeed, it will not occur at all until communities accept a case for change and require policy makers and politicians to assign the requisite funds as a result. As Wegman has pointed out, there is a huge gap between intellectually embracing Safe System principles and applying them as policy into practice.[46]

The most promising development is the movement for widespread adoption of a star rating system for roads analogous to the system developed for cars. The International Road Assessment Program (iRAP) is aligned with the Decade of Action in an attempt to get nations systematically to assess the safety of the road and traffic systems and so prioritise expenditure to greatest effect.[234] Safety stars on cars have become increasingly important in consumer purchasing decisions, and manufacturers have responded by seeking to produce five-star vehicles. Promoting stars on roads may be an effective way of focussing government attention (through community concern) on the more unsafe parts of the road network.

Governments share a reluctance to commit the funds for transformational change, not just because of the need to balance expenditure against perceived priorities, but also because of a reluctance to accept true accountability for system safety for that vast majority of casualty outcomes that do not stem from illegal behaviour. How might the impasse be addressed?

We accept that traffic safety strategies that target continuous reductions in the absolute numbers of serious injuries and deaths are a step forward, but we are critical of the soft targets typically set. Most disturbing, though, is the absence of strategies to implement the Safe System principles over time. We should be *planning* for transformational change.

Planning requires a specific focus on identifying and addressing the barriers to implementation. A suggested approach is to start with the most frequent types of serious casualty crash—in Australia these are the single-vehicle run-off-rural-road crash, intersection crashes, both urban and rural, and crashes involving pedestrians. For each, new design principles can be formulated and costed. Effectiveness and public acceptability could then be tested through the application of demonstration projects which reduces cost to a minimum and allows scientific evidence of impact to be evaluated. This is exactly the approach the Swedes took in confronting the frequency of head-on crashes on much of their rural network. The innovation was to place wire-rope barrier down the centre of an undivided road and provide specifically for overtaking opportunities. At the time the measure was considered impractical, but a demonstration project on a substantial length of rural road virtually eliminated head-on collisions. Moreover, the treatment was cost-beneficial and the investment was far less than conventional road duplication would have been.[235]

The absence of planning for transformational change is a widespread weakness in extant traffic safety strategies. Innovation is vital and can be best achieved through demonstration projects in the short term. Safe System design does not have to be prohibitive in cost.

11 Six Vital Steps toward Zero

11.1 TIME FOR A NEW FOCUS

As we have already pointed out, those working in the policy-making and program administration end of the traffic safety field traditionally considered that their tool kit of countermeasures could be categorised into three sets—education, enforcement, and engineering—collectively known as the 3Es.

Each E was largely the province of one set of players among the myriad institutions whose portfolios impacted on safety outcomes. Over time, sets of players coordinated their efforts in a cooperative manner, sometimes to great effect. For example, random breath testing in Australia required intense police resources and intruded upon innocent drivers, both of which only became publicly acceptable because of intense public education justifying the program to the community and creating a climate of support.[174] Similarly, research into the effectiveness of specific measures under each E has proceeded apace and significant advances have been made; for example, decades of research into a range of enforcement strategies to curb drunk driving have resulted in clear guidelines for effective enforcement practice.[134]

There is absolutely no doubt that we have come a long way in improving the effectiveness of our specific safety measures in each of the fields of education, enforcement, and engineering. The *Safe System* can be considered the modern expression of the 3Es with its cornerstones of safe roads, safe vehicles, safe road users, and safe speeds. It is the breakthrough advance in traffic safety thinking of the last decade, because it focuses on human vulnerability to kinetic energy, accepts the prevalence of human error, and recasts the notion of shared responsibility away from its former primacy on the individual to a primacy on the designers and operators of the road system. Yet, as we stated in our conclusion to Chapter 6, the failure to implement the Safe System is the major barrier to the goal of striving toward zero disabling injury and death.

The problem we have with the 3Es mindset is that the focus remains squarely on the individual measures and not on the bigger picture, on the *what* and not on the *why* or the *how*. We continuously improve our measures by questioning the effectiveness of each. For example, what can we do to change driver attitudes (education), to increase compliance with speed limits (enforcement), or to improve vehicle crashworthiness in side impacts (engineering)? Every piece of evaluation research is important, but of greater importance to us are the broader questions: Where are we trying to get to? How might we get there? The Safe System integrates the Es in a way not previously achieved but is still concerned (mostly) with the what. Science has given us the what in great measure, and that evidence base will continue to build. That is not our concern.

We have argued from the outset that the only acceptable long-term objective is the elimination of disabling injury and death from everyday road use. We concluded Chapter 3 with the statement that the elimination of disabling injury and death had to become a nonnegotiable given in all public policy discussions about road transport. We are clearly nowhere near that point yet. We have shown (see Table 3.1 in Chapter 3) that current traffic safety strategies in motorised countries (implicitly) *plan to accept* large numbers of disabling injuries and deaths during their terms of currency. We do not believe that this is acceptable. This belief is the *why* of the case for transformational change. We must now address the *how*.

We present the how under six headings, each a precondition for achieving transformational change. We suggest the new focus should be on the 6Cs and not on the 3Es:

- **Constituency**—A strong public demand for change.
- **Committed leadership**—To help create the constituency but also to respond to it by driving change.
- **Climate of safety**—Changing the climate around road use to resolve the dilemma of the commons will be vital if a constituency is to build.
- **Capacity building**—Professionals and institutions do not yet have the capacity to embrace transformational change.
- **Cooperation and coordination**—Removing silo walls between and within institutions and facilitating integrated efforts among all key players.
- **Courageous patience**—Transformational change will not occur overnight; the journey will be long and difficult, but we must stay the course.

The first three require a major change in the prevailing culture of road use and in the way we view safety and respond to unsafety. Without cultural change we will remain locked into a process of making only incremental improvements. To achieve cultural change, we need to understand how to influence social change at a macro-level and adopt the successful strategies of others.

11.2   CONSTITUENCY

We have earlier presented evidence that there is widespread public ignorance about the extent and widespread long-lasting impacts of disabling injury and death on our roads, that the daily drip, drip of media coverage of dramatic crashes serves only to inoculate the community, and that crashes are portrayed as blameworthy, happening to misbehaving others.

To make matters worse, bureaucrats and politicians routinely tell their communities of their great safety progress, with such statements as "this year has seen the lowest number of road deaths since records were kept," that "the death rate per unit travel (or per head of population) is at a historical low," that their jurisdiction "is a world leader in traffic safety and the envy of others," etc. Each of these statements is supported by reference to the relevant statistics and can be said to be true, but only in a very superficial sense. The enormity of the trauma we tolerate is always left unsaid, and the urgency of road trauma as a public health problem remains unacknowledged. In Australia almost 34,000 people were seriously injured or killed in road crashes in 2010. No other daily activity exacts such a price. Can we really be doing as well as we are told?

New 10-year strategies are paraded by politicians as "leading edge," with targets that continue the downward trend in disabling injury and death. In the case of Australia, what is not said is that *at least a quarter of a million people will be seriously injured or killed in road crashes during the 10-year life of the current national road safety strategy.* Is that really an acceptable price to pay?

What might happen if the community truly understood the level and nature of the road trauma it is (implicitly) being asked to accept? What might happen if the community was asked *explicitly* to accept the level of trauma its governments are planning to accept? The discussion must become open and transparent. The more controversial the issue, the greater is the need for public transparency. As Ian Lowe wrote: "When the people lead the leaders eventually follow."[236]

We believe that a constituency for traffic safety would emerge from an explicit statement by government of the level of trauma officially acceptable. Recall the example we gave of the four occasions when Victoria introduced radically new traffic safety policies; each one followed public concern over an identified spike in deaths (see Table 8.2). It is not that the community does not care; it is more that the community is not told and therefore does not understand.

As we have already seen, it will not be enough to just put the facts into the public arena; the facts must be supported by effective persuasion of the need for, and feasibility of, transformational change. How might persuasion occur?

The presentation of large numbers (such as the quarter of a million cited above) will not suffice alone. The material has to be made personal. If a consumer magazine reports that the Brand X car is the most reliable, you will not buy one if a respected neighbour has had a bad experience with that brand. If you read that smoking causes lung cancer but none of your personal friends who smoke have cancer, you will not give it up. Personal experience far outweighs statistically framed messages.

Traffic safety needs its own vested interest groups to lobby governments and to lead public debate, using personalised messages. Most road crash victims groups tend to confine their efforts to providing counselling and other forms of personal support to those in need. This valuable service must, of course, continue, but if the raft of such groups combined their energies to put their stories into the public arena, a wider impact would be likely. We defy any reader not to be touched by the three stories we presented in Chapter 2 or the impacts on those emergency service personnel on the front line of trauma that we saw in Chapter 7.

Apart from telling deeply moving personal stories at each and every public opportunity, we must combat the "blame the victim" mindset. This can best be done by using the personal stories of blameless victims like Noel, Sam, and Abbey and the frontline folk like Richard and Kate. We must work with the media to ensure coverage of these cases, particularly the follow-up describing the long-term impacts. We must emphasise the personal impacts of the road trauma.

Beyond victims' groups, other coalitions should form. For example, most parents worry about the safety of their children, initially as pedestrians and later as novice drivers. Armed with facts, parents could become an effective lobby. As Danny Dorling has pointed out (in the United Kingdom), about 40% of children aged 5–10 years and about 50% of those aged 6–11 years who die from external (not disease) causes die from road crashes. Moreover, in the age group 20–24 years there are three major external causes of death—road crashes, suicide, and drug/alcohol overdose—with the first two accounting for about equal numbers and well ahead of the third. Yet the alcohol/drug issue gets all the public attention.[52]

Australia has a successful record in attracting volunteers to a range of causes. A pertinent example is the incidence of volunteering to assist (disadvantaged) learner drivers to obtain the required number of hours of supervised driving practice set by regulation before the licence test can be attempted. It is highly likely that a call for volunteers to press for change in traffic safety policy would be successful, especially among those impacted by road trauma, either directly or indirectly.

Identifying and facilitating champions of traffic safety is another effective method. In the 1970s, surgeons in Australia were instrumental in achieving government focus on injury prevention.[106] There are few champions now. If sports leaders, business leaders, professional leaders, and the like, including politicians, were canvassed, it is highly likely that a number who have been personally "touched by the road toll" would be identified, and many of these would be willing to act as champions in the public debate, again armed with knowledge of the kind we have put forward. Adopting "stars on roads" and arming the champions with powerful data on the unsafety of our infrastructure may be a way of getting governments to do more than conceptually accept Safe System principles.

As public interest builds, we might borrow the concept of powerful symbols from other health causes. If breast cancer prevention advocates can get people to wear a pink ribbon, and if advocates for sudden infant death syndrome (SIDS) can get people to wear large red noses (including noses for their cars!), should we not generate a symbol for road trauma prevention? There is a movement in Australia called Fatality Free Friday, but it gains little traction as yet. We must address the reasons that there is complacency about the extent of road trauma before we implement well-meaning but ineffective symbols.

In Australia, at present there are few effective traffic safety advocates. Who might facilitate the creation of a constituency, a constituency that demands a new approach? It clearly will not be the government! Nor can the front line come from the bulk of traffic safety professionals, as they are employed by the government. The role of academics is moot, as most depend on funds from the government transport institutions for their research; nevertheless, we must seek better links between the ivory towers and the corridors of power.

Existing nongovernment bodies like the Australasian College of Road Safety and 33,900—the Australian Road Safety Collaboration should be the first port of call to initiate and facilitate a strategically planned campaign to create a vibrant constituency, using (at least) all the avenues canvassed above. Enlisting the support of one or more experts in social marketing and lobbying would be extremely beneficial. Public support for water conservation practices and for rubbish recycling are key success stories. We need to examine these and similar case studies to identify the lessons we may adapt to the process of generating a constituency for traffic safety.

11.3   COMMITTED LEADERSHIP

While the concept of shared responsibility is sound, in practice it can prove a convenient place to hide. Shared responsibility should mean exactly that, various parties accepting accountability for achieving the outcomes assigned to them in a collaborative process. However, for decades, it has been used by government agencies to deflect accountability for safety onto the road user, who is deemed to have to *share responsibility*. And so the road user must, but he or she should only be held accountable for behaving in accord with the laws, not for making very human errors in a system that is not error tolerant.

Similarly, in any given jurisdiction, several agencies share responsibility for the overall safety outcome, but rarely are individual agency outcomes specified, so blame for a failure to achieve jurisdictional targets can always be laid at another agency's door. At the national traffic safety strategy level in a federation, shared responsibility means consensus decisions on targets and initiatives. As Osborne and Gaebler, in a telling book titled *Reinventing Government*, said, reform by consensus means reform at the lowest common denominator.[237]

Shared responsibility does not absolve individuals, agencies, and governments from exercising leadership. It is rare in a traffic safety strategy at either the state or national level in Australia to define specific agency accountabilities for outcomes—just who should achieve what by when. The head of each agency with a role to play should grasp the nettle and clearly define the agency's role and its measures for judging outcome achievement. Committed leadership requires nothing less.

Politicians must also demonstrate leadership. Of course, we need first to get them to accept that road trauma is a high-priority public health problem that requires a transformational change of policy! No doubt there are politicians who have been touched by the road toll and can agitate for change within party rooms, within cabinet meetings, or on the floor of parliament. First, though, they need to accept the problem, and a vibrant constituency will no doubt help here.

Leadership is not confined only to those with a direct responsibility for all or part of traffic safety policy. A former French president was mocked during a European heads of government meeting over France's (then) poor traffic safety record. He returned home and instructed the relevant ministers—who in turn instructed the relevant agency heads—to "fix the problem." Within 3 years the number of road crash deaths on French roads had fallen dramatically.[32,238] That is leadership, and the motive, though misplaced, was effective.

In most areas of public policy formulation, but particularly in traffic safety in Australia, a mantra of consultation is chanted. Not only is traffic safety seen as a shared responsibility, but also policy creation is seen to require community involvement. As we have already seen, the public has no real understanding of the state of traffic safety. The popular beliefs about cause and solutions are often wrong, and the myth of blameworthy behaviour holds sway. There is, for example, widespread support for more school-based safety education to generate better attitudes and little support for the development of an error-tolerant road system. This plays right into the hands of those who do not want transformational change. While consultation is important to garner acceptance of proposed policies and programs, it is not a substitute for committed leadership. If the public is ignorant of the true state of affairs, strong leadership is vital to ensure that the knowledge deficit is corrected. If the public comes to understand the problem but does not accept the proposed solution, then that is its right in a democracy. At present, it is being consulted with only partial disclosure of the facts (Box 11.1).

Leadership can exist at many levels. Champions of a cause are leaders—they agitate and promote and are not discouraged easily. Goldstein and his colleagues identified six universal principles of social influence, two of which were *liking* and *social proof*.[173] The more we like someone, the more we accept their views on issues, while the more we admire someone, the stronger the social proof contained in their views. We should seek champions who are widely liked and admired in our community. Identifying popular, admired leaders committed to agitate for change in

---

BOX 11.1   CONTROVERSIAL INTERVENTIONS

Government knows best. Don't share ideas with the public that may upset someone! If you do, rule out in advance those that may be controversial.

Often the views of a minister or cabinet or chief minister are not reflective of an informed community position. Politicians may be reluctant to openly provide information to the community about a particular issue.

This creates considerable difficulty. There is little chance to inform the community of relevant evidence when information provision is constrained in this way.

In a particularly interesting example from Victoria, Australia, information about potential initiatives was provided to the community for discussion, but with a clear accompanying statement that certain of the initiatives were not favoured by government.

Proposals to upgrade graduated licensing arrangements in Victoria were developed by the road safety agencies in the mid-2000s and publicly released.* It outlined the current provisions, the crash risks of certain existing behaviours, the evidence for potential new conditions, and then a proposal for a new graduated licensing system (GLS).

The paper indicated that research had clearly shown that driving late at night, with multiple passengers, drunk driving, using mobile telephones, and towing are much higher-risk activities for young inexperienced drivers, and that half of first-year drivers' fatal crashes occur while driving between 10 p.m. and 6 a.m. or with multiple passengers.

The government required the document to outline which new measures were in fact favoured by it.

This had the effect of ruling out certain options with proven road safety benefits. The discussion paper included the following:

The Government has considered other possible restrictions for P1 licence holders such as a passenger limit and a late night driving restriction (with exemptions).

traffic safety is a high priority. Agency and political leaders are likely to follow such role models.

Traffic safety scientists must also become leaders. We need to learn to communicate more simply and effectively, something most of us are very poor at. The third of the universal principles of social influence is *authority*; we respect experts who can talk at our level. Of course, in any controversial public debate (the strict enforcement of speed limits, for example) every group with a vested interest produces its own "experts," and the result is public confusion. Through a range of professional associations we should seek to present a common scientific message, which means that we should seek to resolve issues of scientific dispute within the professions, rather than in the public arena. We need both accuracy and consistency from the respected professional groups. As the authors of the Organisation for Economic Cooperation and Development (OECD) report on achieving ambitious road safety targets wrote:

> Whilst these restrictions have resulted in road safety benefits in Canada, the US and New Zealand, the Government does not support them at this stage because of concerns about the inconvenience and the impact on young people's social mobility associated with such restrictions. The advantages and disadvantages of such restrictions will require detailed consideration and considerable community debate.
>
> So the paper, in proposing new measures, did not include passenger restrictions or late-night driving restrictions for first-year drivers. They were not favoured by the government of the day.
>
> There was good community support for 120 hours of supervised on-road practice, a new licence test to check whether applicants had in fact received 120 or so hours of on-road practice, and a 1-year plus 3-year probationary licence period (up from 3 years). There was also little negative reaction about passenger restrictions. However, the government continued to refuse to include passenger restrictions in the new proposals.
>
> At the time legislation went forward to adopt the new provisions, there was a change of chief minister. The new chief minister was convinced of the merits of passenger restrictions and they were included at the last minute—perhaps influenced by a (then) recent dramatic crash widely reported in the media involving several teenagers in one car.
>
> Initial evaluation of the new GLS in 2012 has demonstrated substantial casualty crash reduction benefits (some 30%). While the contribution of passenger restrictions cannot be singled out, the research is clear about its safety benefits.
>
> It nearly did not happen due to a reluctance by government at the time to give the community a chance to pursue it, even though information on the implementation and benefits of the proposal (including likely fatality and serious injury reductions) had been provided and there had not been a negative public response.

*Victorian Government, Young Driver Safety and Graduated Licensing, Discussion Paper, August 2005.

"Road safety practitioners and stakeholders have a responsibility to influence the political process of policy assessment through competent and persistent advocacy through government."[135]

Where traffic safety professionals can exercise leadership without coming into conflict with their government employers is in the formulation of a phased strategy for the implementation of Safe System principles, demonstrating what could be achieved in a staged process, at what cost, and by whom. For the design of new roads, the road safety audit processes in widespread use in many motorised nations are valuable tools, but designers need to go much further. They need to consider the complete operating environment of the road—how to meet the needs of cyclists and pedestrians, how to design error-tolerant intersections. Some road authorities have taken this broader approach, but it requires senior executive leadership for its initiation. Adopting stars on roads as a simple means of both communication with the community and prioritising investment would be a leap forward.

There has been a welcome shift from the retrospective black spot engineering approach—where treatments are applied to sites (or lengths of road) with high casualty crash numbers—to the prospective approach of undertaking traffic safety risk assessments to identify sections with design, construction, or traffic management issues that may increase future crash probabilities.

What we are suggesting here is a further development to expand the proactive approach to design in order to manage crash risk. One example from Victoria will suffice to illustrate the principle. Single-vehicle run-off-road crashes represent the single most frequent crash type on Victorian rural roads. Because they do not cluster at specific spots or even lengths of road, it is difficult to apply retrospective treatments. At the same time, the funds are not available for universal prospective treatments. Under a Safe System strategy the roadsides would be cleared along their entire length or the roadside obstacles would be protected by barriers or the speed limit would be reduced to a level where crashes of this type would be unlikely to result in serious injury. A staged implementation could involve placing barriers along, say, the first 50 kilometres of each of the five major highways leading out of Melbourne, with speed limit reductions on the rest, with those reductions gradually being removed as funds become available to extend the barrier implementation. The (arbitrary) 50 kilometre starting radius covers those stretches of road with the highest traffic volumes, and therefore the highest risk of a high absolute number of crashes.

11.4 CLIMATE OF SAFETY

The fourth of Goldstein and colleagues' universal principles of social influence is the *need to behave in ways that are consistent with our values*. We breach this principle constantly in traffic safety policy and practice. We know, for example, how critical speed moderation is, but we behave as though it is of very low importance. There is not one vehicle design rule dealing with top speed capability or acceleration performance; the design of the speedometer is not fit for purpose; vehicle advertising that promotes speed continues largely unabated; politicians publicly decry intense enforcement of speed limits whenever it suits them; and speed limits are typically set to maximise traffic flow and are rarely matched with the level of safety provided by the road section in question.

We have explored at some length why it is that road users choose behaviours with instant personal rewards over those that would improve safety for the wider community. Why should they do otherwise when official policies are at such odds with a purported concern with safety?

How might we commence a process through which a climate of safety might emerge around our daily road use? The first step is to identify and remove all these inconsistencies. In the case of speed moderation, we need to open a discussion with the vehicle manufacturers and their advertisers, not to declare war. How might sales volumes be maintained or increased using other sales pitches? Governments could fund research and trials, far preferable to attempting to regulate through design or advertising standards. The saviour may well be technological, where a smart road advises the vehicle of the prevailing speed limit and the smart vehicle limits its speed to that level. Right now, intelligent speed assist (ISA) does this in an advisory rather than a controlling manner. An open dialogue between government and industry with incentives to change is likely to prove a powerful change agent. Sadly, such a dialogue has yet to begin.

The fifth of Goldstein and colleagues' universal principles is *reciprocation*—we return favours to those who favour us. If we were to portray nonspeeders as excellent citizens and great role models to their children and peers, we are likely to have a stronger influence on speeding than if we continue to focus on speeders as bad people. Currently, we report the number of speeding tickets issued in a police blitz rather than reporting the incredibly high proportion of drivers who pass speed camera sites behaving safely. We need to switch our messages from the negative to the positive. Wouldn't it be great if the frequency of cheery waves acknowledging courteous behaviour outweighed "the bird," which typifies the mild end of road rage?

We have grounds for real optimism. The climate around smoking has changed to a degree where smoking around others (offices, restaurants, sporting venues, public transport, etc.) is no longer acceptable. Smokers can no longer consider only their own desires. While we have a very long way to go to reach an analogous point with drivers, the journey must begin in earnest.

Our optimism is also based on Australia's fine record of leading public perception and opinion through government action to introduce world-leading traffic safety behavioural measures. Mandatory seat belt wearing, random breath testing for alcohol, automated speed enforcement, random roadside saliva testing for drugs, alcohol interlock schemes for drunk driving offenders, graduated licencing regimes for novice drivers, and other strategies were all contentious at the time of introduction, but are now accepted and are credited with dramatically reducing road crash deaths.

When advertisers sell junk food, we expect parents to exercise responsibility. When the poor are seduced by subprime mortgages, we expect them to exercise responsibility. When unsolicited offers come through the mail to increase our credit card limit, we are supposed to exercise responsibility. This is the standard argument of the "nanny staters"—everyone should be free to take their own risks. Tim Wilson from Australia's Institute of Public Affairs wrote a commentary article in *The Age* in March 2012. In part, he wrote: "No-one disputes that free people can make bad decisions. Without our mistakes we cannot learn. It's immoral to create a risk-free society that infantilises people.... Nanny's supporters too often confuse their own subjective views with what is objectively right."[239] The flaw in the argument is that driving is not a private activity where the risk is purely personal. Others are put at risk of serious injury or even death.

The nanny staters chant the mantra of personal responsibility, but personal responsibility means more than considering only personal outcomes. Even in a single-vehicle crash the community bears disproportionate costs. If a motorcyclist, for example, wishes to refrain from wearing a helmet, he should either sign a waiver to forgo publicly funded treatment in the event of a head injury from a motorcycle crash or pay an insurance premium commensurate with the likely cost of treatment. When exposure to risk is collective, there is a collective responsibility that cannot be abrogated.

The Dutch are providing a lead with research into what they call social forgivingness, the development of a cooperative rather than a competitive climate around road use. The term we might use is *duty of care*. It will not be easy, and it will not be short term, but we must develop a strategy specific to this objective. The principles of preventive medicine presented in Chapter 8 are principles involving preventive action at the community level. There are case studies from public health that we can emulate.

As Vanderbilt points out, our driving practices are based on personal assumptions.[60] We race to and through intersections on the assumption that no one will run the red (or the stop sign) from the crossroad, and we run the red very early in the cycle because we assume no one will "jump" early from the crossroad. Ramp metering on entrances to motorways is an example of a traffic engineering tool that promotes cooperation, albeit through regulation. What other traffic engineering designs to facilitate cooperation might we develop?

A case exists for intensive public education campaigns, such as are commonly publicly funded in Australia, to promote specific, cooperative behaviours. They are only likely to be considered and funded if the need for the creation of a climate for safe road use is accepted, and if a specific strategy for its long-term implementation is developed.

Perhaps one of the most powerful ways to create a climate of safety around road use will prove to be leadership through role modelling from corporations, both private and public. Occupational safety is, in most advanced societies, light years ahead of traffic safety in its focus on safe deign, safe operations, and safe cultures. It can be argued that this is because industry is a closed-loop system in that the employer has control over all system elements, including over employees' safety-related actions and behaviours. It is worth noting that governments made employers accountable for industrial safety but seem reluctant to accept a similar accountability themselves for the system they own and operate.

Corporations have come to recognise that road crashes represent the single largest cause of workplace death. Beyond the obvious industrial segment of professional driving (buses, taxis, couriers, freight transporters, etc.), crashes occur to tradesmen moving from job to job, sales representatives moving from client to client, office workers moving from meeting to meeting, and others during their daily commuting journey. In addition, there are crashes on mine sites, in warehouse yards, and so on.

In Australia, it has been estimated that work-related road crashes account for about 50% of occupational deaths and about 15% of all road deaths.[186] Many companies are responding with significant strategies and action plans.

One of the world's largest companies—BHP Billiton—has implemented a policy to purchase only workhorse vehicles that achieve a five-star safety rating without modification and to require all of its contractors to comply with this standard. This is leading manufacturers to respecify their standard workhorse models. The ongoing effect is to raise the safety standard of all such vehicles in the entire Australian fleet, no matter who buys them.

Safety policies that include banning the use of cell phones while mobile, prescribing zero blood alcohol levels, managing rest breaks to decrease fatigue risks, mandating safety training, and a host of other measures are being introduced in an increasing number of companies. The modelling of a safe road use culture among employees—as well as a safe culture within the confines of the workplace—has enormous potential for a flow-on effect. Employees are likely to apply the same practices in their private driving as they do in their work-related driving. Government departments should join this movement and ensure they also become appropriate role models.

11.5 CAPACITY BUILDING

If the Safe System is to be implemented effectively, it needs to be understood and accepted by frontline staff, not just their managers. Such staff comprise traffic engineers, infrastructure engineers, police, town planners, transport planners, vehicle engineers, educators, and many others across the full range of agencies whose work directly impacts upon traffic safety performance.

In Australia, there is little systematic effort to build capacity, either among the managers of groups whose work is not traffic safety per se, but which influences safety outcomes, or sadly even at the frontline level. The Queensland University of Technology offers a certificate course for practitioners such as local government officers, police, and the like, and a traffic safety leadership short course for senior agency staff is under development as a joint effort between two universities in Victoria and South Australia.[240,241] These are promising beginnings, but much more needs to be done.

In particular, transport agencies responsible for road system design, construction and maintenance, and traffic operations should ensure their staff understand, accept, and apply Safe System principles. In the short term, this will require the conduct of regular in-service training, and in the longer term, the inclusion of appropriate units in undergraduate courses.

The most urgent need is for the development of an explicit strategy to implement Safe System principles throughout all aspects of the planning, design, and operation of the road transport system. At present, implementation relies entirely on a handful of advocates being able to convince others of the value of a change in mindset. As we have seen, such efforts are slow to take root because of a lack of acceptance at the highest levels of leadership. We have national traffic safety strategies notionally built upon Safe System principles, but with no strategy for the implementation of such principles.[242]

Committed leadership involves not only helping build a public constituency for a sea change in effort, but also exerting pressure to build the institutional capacity for such a change to be realised.

11.6   COOPERATION AND COORDINATION

Australia has been recognised internationally in traffic safety primarily for its innovative approach to regulation and its intense enforcement supported by extensive public education.[174] What has been at least as important, but has gone largely unrecognised, is the extent to which agencies have cooperated and, with telling effect, coordinated their traffic safety efforts.

Perhaps the best example relates to the success of random breath testing in the state of Victoria. The introduction of a law that permitted police to require a breath test of any driver without cause, including via roadblocks that tested large numbers of drivers at random, was revolutionary. Initially it was also ineffective. Research was commissioned to try to understand why it did not work. The police cooperated with the academics and reassigned enforcement resources according to a plan to scientifically assess the effect of enforcement intensity. The data suggested an operational rule of thumb of 20 hours of enforcement per 100 square kilometres per week to achieve effectiveness.[243]

The police lacked both the equipment and manpower resources to deliver this enforcement level. However, the (government) injury compensation insurer saw, from the science, potential cost savings for its business and funded the purchase of the equipment the police needed and met the costs of overtime payments to enable the police to meet the manpower costs. In return, it required the enforcement effort to comply with the methods recommended by the research findings. The insurer also funded massive public education campaigns to generate public support for this uniquely intense enforcement effort. Within 3 years the proportion of fatally injured drivers with blood alcohol levels above the legal limit had fallen from around 50% to around 20%.[120,243]

There is a very powerful lesson here of the benefits that follow from planning traffic safety efforts to maximise the synergies between the efforts of agencies that traditionally operate independently of one another.

Three countries with the best safety records—Sweden, the United Kingdom, and the Netherlands—initiated a research project to seek an explanation for their success. They discovered that each had achieved safe outcomes down different paths, that the answer lay more in how they went about their task than in the particular package of measures they used. They concluded that having good data to define the specific problems, selecting measures of known effectiveness, and focussing on effective implementation were the key ingredients; in short, that excellent management was crucial.[244]

The OECD set up an international expert group to examine how to improve the effectiveness of traffic safety efforts. The report—*Towards Zero: Ambitious Road Safety Targets and the Safe System Approach*—was released in 2008. While it contains a useful overview of effective safety measures, its major contribution is its focus on how to manage traffic safety efforts for maximum synergies. It concluded: "Consideration of all elements of the road safety management system becomes critical for any country seeking to surpass its current performance levels and to go beyond good practice outcomes to achieve even more ambitious results" and "[the shift to a Safe System approach] requires a strong commitment to institutional capacity building and ongoing innovation, sustained by the process of research and knowledge transfer across international boundaries."[245]

The importance of effective management across the disparate agencies whose policies and practice influence safety outcomes cannot be overemphasised. Capacity building, coordination, and cooperation are paramount to success.

11.7 COURAGEOUS PATIENCE

In 1997 the Swedish parliament had the courage to pass a resolution to the effect that no one behaving reasonably should be seriously injured or killed through their daily use of the roads. The Swedish Vision Zero drew a line in the sand. Nonetheless, more than 15 years later, the advocates of this vision still encounter opposition based on questions of practicality and cost.[61] Having a vision is one thing; realising it is quite another.

We have the knowledge to achieve safe road transport use, and this knowledge grows almost daily. What we do not have is the willingness to implement this knowledge to its full. Until we do, the best we can hope to achieve is a continuing but far too gradual decrease in the number of disabling injuries and death.

The future does not arrive on the doorstep unannounced. To make a quantum leap forward, we need to create a constituency for safety; we need committed leadership; we need to build a professional and practitioner capacity for transformational change; we need to engender a climate of cooperative road use as a path to a safe culture; and we need cooperation and coordination as hallmarks of effective management of traffic safety policy and practice. Each is dependent upon the others; none is sufficient alone. Given everything we understand about the barriers to transformational change, we need courageous patience. We must not get discouraged on our journey. It will be hard but we must persist.

References

1. WHO. *Global status report on road safety: Time for action.* 2009, World Health Organization, Geneva.
2. The Lancet. Global Burden of Disease Study 2010. *Lancet*, 2012, 380(9859).
3. Mohan, D. Road traffic injuries: A neglected pandemic. *Bulletin of the World Health Organisation*, 2003, 81(9): 684–685.
4. Marr, D. *Panic.* 2011, Black, Inc., Melbourne, Australia.
5. Evans, L. *Traffic safety.* 2004, Science Serving Society, Bloomfield Hills, MI, p. 1.
6. Nevada DOT. *Zero fatalities NV: What should be our goal?* 2012.
7. Guggenheim, D. *An inconvenient truth.* 2006, Paramount Classics.
8. Williams, A. Reflections on the highway safety field. In *Patricia F. Waller memorial lecture.* 2004, Highway Safety Research Center, University of North Carolina, Chapel Hill, NC.
9. Williams, A., and N. Haworth. Overcoming barriers to creating a well-functioning safety system: A comparison of Australia and the United States. In *Improving traffic safety culture in the United States: The journey forward*, AA Foundation of Traffic Safety. 2007, Washington, DC.
10. Roberts, I., and K. Abbasi. Editorial—Death on the roads: Two years on. *British Medical Journal*, 2004, 328(7444): 845.
11. Allsop, R. Risk of road use as a daily activity. Personal communication with Ian Johnston at the Scientific Research on Road Safety Management Workshop, Haarlem, Netherlands, November 2009.
12. Levitt, S.D., and S.J. Dubner. *Superfreakonomics.* 2009, William Morrow, New York.
13. Shinar, D. *Traffic safety and human behavior.* 2007, Elsevier, Amsterdam.
14. Elvik, R., T. Vaa, and A. Erke. *The handbook of road safety measures*, 2nd ed. 2009, Elsevier, Amsterdam.
15. ATSB. *Characteristics of fatal road crashes during national holiday periods.* 2006, Australian Transport Safety Period, Canberra, ACT.
16. Schmich, M. Advice, like youth, probably just wasted on the young. *Chicago Tribune*, 1997.
17. WHO. *Global status report on road safety: Time for action.* 2012, World Health Organization Department of Violence and Injury Prevention and Diability, Geneva, Switzerland.
18. Worley, H. *Road traffic accidents increase dramatically worldwide.* 2006, Population Reference Bureau, Washington, DC.
19. WHO. *World report on road traffic injury prevention*, ed. R.S.M. Peden, D. Sleet, D. Mohan, A.A. Hyder, E. Jarawan, and C. Mathers. 2004, World Health Organization, Geneva.
20. Make Roads Safe Foundation. *The campaign for global road safety.* 2012. http://www.makeroadssafe.org/Pages/home.aspx.
21. UN. Moscow Declaration. In *First Global Minesterial Conference on Road Safety: Time for action.* 2009, Russian Federation, Moscow.
22. Kelly, A. Roads still unsafe because of 'miserably inadequate' funding: Road safety doesn't have political or financial support to move it into mainstream of development work, says new report. *Guardian*, 2013.
23. Bliss, T., and J. Breen. *Implementing the recommendations of the World Report on Traffic Injury Prevention.* 2009, World Bank, Washington, DC.
24. PIARC. *Comparison of national road safety policies and plans.* 2012, PIARC Technical Committee C.2 Safer Road Operations.
25. Staley, S., and A. Moore. *Mobility first.* 2009, Rowman & Littlefield, Lanham, MD.

26. Johnston, I., et al. *Reducing serious injury and death from run-off-road crashes in Victoria—Turning knowledge into action.* Monash University Accident Research Centre Report Series. 2005, Monash University Accident Research Centre, Melbourne, Australia.
27. Hansson, S.O. Safety is an inherently inconsistent concept. *Safety Science*, 2012, 50(7): 1522–1527.
28. Ministry of Transport and Communications. *En route to a society with safe road traffic: Memorandum.* 1997, Swedish Ministry of Transport and Communications, Sweden.
29. ATC. *National road safety strategy 2011–2020.* 2011, Australian Transport Council, Canberra.
30. Johnson, R. Address to parliament: Road safety strategy to reduce road trauma in Western Australia 2008–2020. In *Towards zero—Road safety strategy.* Officially endorsed by government March 2009.
31. AAA Foundation for Traffic Safety Research. *Improving traffic safety culture in the United States: The journey forward.* 2007, AAA, Washington, DC.
32. Johnston, I. Beyond 'best practice' road safety thinking and systems management: A case for culture change research. *Safety Science*, 2010, 48: 1175–1181.
33. Trinca, G.W., et al. *Reducing traffic injury—A global challenge.* 1988, Royal Australasian College of Surgeons, Melbourne.
34. Reurings, M.C.B., et al. *Why do the number of serious road injuries and the development of the number of road fatalities differ? An analysis of the differences between the developments.* 2012, SWOV, Leidschendam, Netherlands.
35. The Australian Road Safety Collaboration. *33,900.* 2013. www.33900.org.au.
36. IRTAD. *IRTAD road safety 2011 annual report*, OECD. 2012, International Traffic Safety Data and Analysis Group, Paris.
37. MUARC. Unpublished data: Serious injury and severity rates. 2012, Monash University Accident Research Centre, Melbourne, Australia.
38. IRTAD. *IRTAD road safety 2010 annual report*, OECD. 2011, International Traffic Safety Data and Analysis Group, Paris.
39. Smeed, R.J. Some statistical aspects of road safety research. *Journal of the Royal Statistical Society: Series A (General)*, 1949, 112(1): 1–34.
40. IRTAD. *IRTAD annual report 2010.* 2010, International Traffic Safety Data and Analysis Group.
41. Koptis, E., and M. Cropper. *Traffic fatalities and economic growth.* Policy Research Working Paper 3035. 2003, World Bank, Washington, DC.
42. Mathers, C., and D. Loncar. *WHO: Updated projections of global mortality and burden of disease.* 2005.
43. NHTSA. *2010 overview traffic safety fact sheet.* 2012, National Highway Traffic Safety Administration, Washington, DC.
44. Elvik, R. Can injury prevention efforts go too far? Reflections on some possible implications of Vision Zero for road accident fatalities. *Accident Analysis and Prevention*, 1999, 31: 265–286.
45. Allsop, R., N. Sze, and S. Wong. An update on the association between setting quantified road safety targets and road fatality reduction. *Accident Analysis and Prevention*, 2011, 43: 1279–1283.
46. Wegman, F., and S. Oppe. Benchmarking road safety performance of countries. *Safety Science*, 2010, 48: 1203–1211.
47. UN. *Human development report 2011.* 2011, United Nations Development Programme, New York.
48. Tibbs, H. Factors shaping the future of the transport system. In *Proceedings of the 19th ARRB Conference.* 1998, ARRB Group, Sydney, Australia.
49. Featherstone, M. Automobiles: An introduction. *Theory, Culture and Society*, 2004, 21: 1–24.

50. Redshaw, S., and G. Noble. *Mobility, gender and young drivers: Second report of the Transforming Drivers Study*. 2006, University of Western Sydney, Sydney, Australia.
51. Davison, G. *Car wars: How the car won our hearts and conquered our cities*. 2004, Allen & Unwin, Sydney, Australia.
52. Dorling, D. *The 21st Westminster lecture on transport safety*. 2011, Parliamentary Advisory Council for Transport Safety, European Transport Safety Council, London.
53. Miller, D. *Driven societies: Car cultures*, ed. D. Miller. 2001, Berg, Oxford.
54. Beckman, J. Mobility and safety. *Theory, Culture and Society*, 2004, 21: 81–100.
55. Urry, J. *Sociology beyond societies*. 2000, Routledge, London.
56. Newman, P., and J. Kenworthy. *Sustainability and cities: Overcoming automobile dependence*. 1999, Island Press, Washington, DC.
57. Elvik, R. The mistaken belief in the incentives generated by cost-benefit analysis. In *Scientific research on road safety management: Workshop*. 2009, SWOV, Leidschendam, Netherlands.
58. Watt, N. David Cameron unveils plan to sell off the roads. *Guardian*, March 19, 2012.
59. AAA. *2011 Traffic Safety Culture Index*. 2012, American Automobile Association Foundation for Traffic Safety, Washington, DC.
60. Vanderbilt, T. *Traffic: Why we drive the way we do (and what it says about us)*. 2008, Vintage Books (Random House), New York.
61. Tingvall, C. The Swedish position. Personal communication with Ian Johnston. 2012.
62. NTC. *Performance based standards regulation impact statement March 2012*. 2012, National Transport Commission, Melbourne, Australia.
63. RACV. *RACV directions: What's important to our members*. 2007, Royal Automobile Club Victoria, Melbourne, Australia.
64. Moriarty, P. Changing the car culture. In *Low carbon transport for our cities*. 2008, University of Melbourne, Melbourne, Australia.
65. Cooke, C., and J. Puddifoot. Gun culture and symbolism among UK and US women. *Journal of Social Psychology*, 2000, 140(4): 423–433.
66. Hirst, J. *The Australians: Insiders and outsiders on the national characters since 1770*. 2007, Black, Inc., Melbourne, Australia.
67. Pappas, S. Cars with big grilles look like old men. *LiveScience*, 2011.
68. Windhager, S., et al. Face to face: The perception of automotive designs. *Human Nature*, 2008, 19(4): 331–346.
69. Windhager, S., et al. Cars have their own faces: Cross-cultural ratings of car shapes in biological (stereotypical) terms. *Evolution and Human Behaviour*, 2012, 33(2): 109–120.
70. Courtenay, W. Constructions of masculinity and their influence on men's well-being: A theory of gender and health. *Social Science and Medicine*, 2000, 50: 1385–1401.
71. Graham, H., and R. White. Young people, dangerous driving and car culture. *Youth Studies Australia*, 2007, 26(3): 28–35.
72. Palk, G., T. Cocker, and J. Freeman. Women, drinking and aggression on a night out: Do girls just wanna have fun? In *APS Forensic Psychology National Conference*. 2011, Noosa, Queensland.
73. Armstrong, K., and D. Steinhardt. Understanding street racing and 'hoon' culture: An exploratory investigation of perceptions and experiences. *Journal of the Australasian College of Road Safety*, 2006, 17(1): 38–44.
74. Dingus, T., et al. *The 100-car naturalistic driving study. Phase II: Results of the 100 car field experiment*. 2006, Virginia Tech Transportation Institute.
75. Rajan, S. *The enigma of automobility*. 1996, University of Pittsburgh Press, Pittsburgh, PA.
76. Jiggins, S. *The 2008 Churchill Fellowship to study media reporting of road crashes*. 2008, Australasian College of Road Safety, Canberra.
77. Girasek, D. Moving America towards evidence-based approaches to traffic safety. In *Improving traffic safety culture in the United States: The journey forward*. 2007, AAA Foundation for Traffic Safety, Washington, DC.

78. Milne, P.W. *Fitting and wearing of seat belts in Australia: The history of a successful countermeasure*. 1985, Australian Government Publishing Service, Canberra.
79. Wilde, G. The theory of risk homeostasis: Implications for safety and health. *Risk Analysis*, 1982, 2(4): 209–225.
80. Adams, J. *Risk*. 1995, Routledge, London.
81. Shinar, D. *Traffic safety and human behaviour*. 2007, Elsevier, Amsterdam.
82. Petras, M. *Why don't they call me?* 2011, CreateSpace.
83. Centre for Advanced Journalism. *Black Saturday: How the media covered Australia's worst peace-time disaster*. 2009, University of Melbourne, Melbourne, Australia.
84. Muller, D., and M. Gawenda. Ethical free-for-all over media access to the fire zone. *Media International Australia*, 2010, 137: 71–79.
85. Dann, S., and M.-L. Fry. Benchmarking road safety success: Issues to consider. *Australian Marketing Journal*, 2009, 17: 226–231.
86. Martin, J., K. Smith, and M. Worth. Aspects of meaning and relevance in news media coverage of motor vehicle accidents. In *Improving traffic safety culture in the United States: The journey forward*. 2007, AAA Foundation for Traffic Safety, Washington, DC, pp. 273–280.
87. Heng, K., and A. Vasu. Newspaper media reporting of motor vehicle crashes in Singapore: An opportunity lost for injury prevention education? *Journal of Emergency Medicine*, 2010, 17: 173–176.
88. Wundersitz, L. *An analysis of young drivers in crashes using in-depth crash investigation data*. 2012, Centre for Automotive Research, University of Adelaide, Adelaide, Australia.
89. MacRitchie, V., and M. Seedat. Headlines and discourses in newspaper reports on traffic accidents. *South African Journal of Psychology*, 2008, 38(2): 337–354.
90. Shinar, D. Aggressive driving: The contribution of drivers and the situation. *Transportation Research Part F: Traffic Psychology and Behaviour*, 1998, 1(2): 137–160.
91. Jones, B. Intelligent discussion all but extinct: People are better educated than ever yet debate is dumbed down. *The Age*, 2011.
92. *Urban Dictionary*. Anecdata. 2012.
93. Cary, T., and N. Wilson. Lewis Hamilton claims Vic Police 'loving' his hoon charge. *Herald Sun*, 2010.
94. Griffiths, J. Sorting the Galileos from the heretics on climate change. Commentary in *The Sunday Age*. 2012, Melbourne, Australia.
95. Buckingham, A. Speed traps: Saving lives or revenue raising? *Policy*, 2003, 19(3): 3–12.
96. Cameron, M., and A. Buckingham. Speed off (Debate). *Policy*, 2003–04, 19(4): 60–64.
97. Global Road Safety Partnership. *Global Road Safety Partnership Newsletter*. 2012.
98. Clark, J. Road safety and the historical perspective. *Roadwise*, 2002, 13(4).
99. Wells, H. *The fast and the furious: Drivers, speed cameras and control in a risk society*. 2011, Ashgate Publishing Group, Surrey, UK.
100. Tillman, W., and G. Hobbs. The accident-prone automobile driver. *American Journal of Psychiatry*, 1949, 106: 321–331.
101. Whitlock, F. *Death on the road: A study in societal violence*. 1971, Tavistock, London.
102. Reason, J. *Human error*. 1990, Cambridge University Press, Cambridge.
103. Nader, R. *Unsafe at any speed*. 1965, Grossman Publishers, New York.
104. Haddon, W.J. Options of the prevention of motor vehicle crash injury: Keynote address to the Conference on the Prevention of Motor Vehicle Crash Injury. *Israel Journal of Medical Sciences*, 1980, 16(1): 45–68.
105. Delaney, A., K. Diamantopolou, and M. Cameron. *MUARC's speed enforcement research: Principles learnt and implications for practice*. 2003, Monash University Accident Research Centre, Melbourne, Australia.
106. Birrell, J. *Twenty years as a police surgeon*. 2004, Brolga Publishing, Melbourne, Australia.

107. Johnston, I. Will a 4WD strategy work in the shifting sands of policy? *Road and Transport Research*, 2002, 11(1): 66–72.
108. Lane, J. Safety in transport. In *Special lectures in transport: National transport policy*. 1968, University of Melbourne, Melbourne, Australia.
109. Wegman, F. *Driving down the road toll by building a safe system*. 2012, Government of South Australia.
110. Media Monitors. 3AW broadcast: Victoria government's strategy to reduce the road toll: Interview with the Victorian premier. ID M000005562741. 2001.
111. Goh, Y.M., H. Brown, and J. Spickett. Applying systems thinking concepts in the analysis of major incidents and safety culture. *Safety Science*, 2010, 48: 302–309.
112. McKenna, F. The perceived legitimacy of intervention: A key feature for road safety. In *Improving traffic safety culture in the United States: The journey forward*. 2007, AAA Foundation for Traffic Safety, Washington, DC.
113. McClure, R., M. Stevenson, and S. McEvoy. *The scientific basis of injury prevention and control*. 2004, IP Communications, Melbourne.
114. Johnston, I., and M. Cameron. *The use of television publicity to modify seat-belt wearing behaviour*. 1979, Federal Office of Road Safety, Canberra, Australia.
115. Johnston, I. Traffic safety education: Panacea, prophylactic or placebo? *World Journal of Surgery*, 1992, 16(3): 374–378.
116. Australian Government. Fatal road crashes in Australia in the 1990s and 2000s: Crash types and major factors. In *Bitre information sheet*, Department of Infrastructure and Transport, Transport and Regional Economics. 2011.
117. Hauer, E. *Challenging the old order: Towards new directions in traffic safety theory*. 1990, Transactions, New Brunswick, NJ.
118. Wundersitz, L., and M. Baldock. *The relative contribution of system failures and extreme behaviour in South Australian crashes*. 2011, Centre for Automotive Safety Research, Adelaide, Australia.
119. Johnston, I.R. Highway safety. In *The handbook of highway engineering*, ed. T. Fwa. 2006, CRC Press (Taylor & Francis), New York.
120. Broughton, J. The correlation between motoring and other types of offence. *Accident Analysis and Prevention*, 2007, 39(2): 274–283.
121. Reason, J. Achieving a safe culture: Theory and practice. *Work and Stress*, 1998, 12(3): 293–306.
122. Saad, F. Ergonomics of the driver's interface with the road environment: The contribution of psychological research. In *Human factors for highway engineers*, ed. R. Fuller and J. Santos. 2002, Pergamon, London.
123. Salmon, P., M. Regan, and I. Johnston. *Human error and road transport: Literature review and recommendations for error management*. 2005, Monash University Accident Research Centre, Melbourne, Australia.
124. Amalberti, R. The paradoxes of almost totally safe transportation systems. *Safety Science*, 2001, 37: 109–126.
125. Candappa, N., and B. Corben, *Best practice in intersection layout with roundabout and traffic signal focus: Literature review*. 2006, Monash University Accident Research Centre, Monash University, Melbourne, Australia.
126. BBC. *Yes minister*. 1984, BBC.
127. Hauer, E. The road ahead. *Journal of Transportation Engineering*, 2005, 333–339.
128. Dunn, J. Traffic safety and public policy. In *Challenging the old order—Towards new directions in traffic safety theory*, ed. J.P. Rothe. 1990, Transaction, New Brunswick, NJ.
129. Howard, E. The value of a life. Personal communication with Ian Johnston. 2013.
130. Rose, G., K.-T. Khaw, and M. Marmot. *Rose's strategy of preventive medicine*. 2008, Oxford University Press, Oxford.

131. Koppel, S., et al. How important is vehicle safety in the new vehicle purchase process? *Accident Analysis and Prevention*, 2008, 40(3): 994–1004.

132. Belin, M. *Public road safety policy change and its implementation—Vision Zero, a road safety policy innovation*. 2012, Karolinska University, Sweden.

133. McKay, H. Keynote speech. In *Australasian Road Safety Research, Policing and Education Conference*. 2011, Perth, Australia.

134. Delaney, A., K. Diamantopolou, and M. Cameron. *Strategic principles of drink-driving enforcement*. 2006, Monash University Accident Research Centre, Melbourne, Australia.

135. OECD. *Towards zero: Ambitious road safety targets and the Safe System approach*, ed. International Transport Forum. 2008, Organisation for Economic Cooperation and Development, Paris.

136. Friedman, T.L. Why nations fail. *New York Times Sunday Review*, 2012.

137. Acemoglu, D., and J. Robinson. *Why nations fail: The origins of power, prosperity, and poverty*. 2012, Crown Business, New York.

138. Thomas, M. A systematic review of the effectiveness of safety management systems. In *ATSB Transport Safety Report*. 2012, Commonwealth of Australia, Canberra.

139. Bliss, T., and J. Breen. Country guidelines for the conduct of road safety management capacity reviews and the specification of lead agency reforms, investment strategies and Safe System projects. In *Implementing the recommendations of the World Report on Road Traffic Injury Prevention*. 2009, World Bank Global Road Safety Facility, Washington, DC.

140. McKenna, F. *Education in road safety: Are we getting it right?* 2010, RAC Foundation, London.

141. Freeman, J. *Shrinking the world: The 4,000 year story of how email came to rule our lives*. 2009, Penguin Australia, Sydney.

142. Levitt, S.D., and S.J. Dubner. *Freakonomics: A rogue economist explores the hidden side of everything*. 2009, HarperCollins, New York.

143. Johnston, I. *Halving roadway fatalities: A case study from Victoria, Australia, 1989–2004*. 2006, Federal Highway Administration, Washington, DC.

144. Hardin, G. The tragedy of the commons. *Science*, 1968, 162(3859): 1243–1248.

145. Waller, P.F. The highway transportation system as a commons: Implications for risk policy. *Accident Analysis and Prevention*, 1986, 18(5): 417–424.

146. Bentham, J. *A fragment on government*. 1776, London.

147. Perlman, E. *Seven types of ambiguity*. 2003, Faber and Faber, London.

148. Rothwell, J. *In the company of others: An introduction to communication*. 2010, Oxford University Press, New York.

149. NHTSA. Seatbelt use in 2011. In *Traffic safety facts*. 2012, National Highway Traffic Safety Administration, Washington, DC.

150. CARRS-Q. *State of the road: Seat belts*. 2012, Centre for Accident Research and Road Safety, Queensland, Australia.

151. NHTSA. *Seat belt use in 2010—Overall results*. 2010, National Highway Traffic Safety Administration, Washington, DC.

152. ETSC. *Seat belts and child restraints*. 2005, European Transport Safety Council, Brussels, Belgium.

153. Insurance Institute for Highway Safety (IIHS). Highway loss data. 2013. http://www.iihs.org/hldi.

154. Burkeman, O. Deadly riding without helmets. *The Age*, 2006.

155. NHTSA. *Evaluation of the repeal of the all-rider helmet law in Florida*. 2005, National Highway Traffic Safety Administration, Washington, DC.

156. Pickrell, T., and M. Starnes. *An analysis of motorcycle helmet use in fatal crashes*. 2008, Report DOT HS 811 011 NHTSA DOT, Washington, DC.

157. Wigglesworth, E. Do some U.S. states have higher/lower injury mortality rates than others? *Journal of Trauma: Injury, Infection and Critical Care*, 2005, 58(6): 1144–1149.
158. Hemenway, D. *While we were sleeping: Success stories in injury and violence prevention*. 2009, University of California Press, Berkeley, CA.
159. Wundersitz, L., T. Hutchinson, and J. Woolley. *Best practice in road safety mass media campaigns: A literature review*. 2010, Centre for Automotive Safety Research, University of Adelaide, Adelaide, Australia.
160. Rose, G. Strategy of prevention: Lessons from cardiovascular disease. *British Medical Journal*, 1981, 282: 1847–1851.
161. Currie, D. Childhood vaccination rates high, but measles re-emerging. *Nations Health*, 2008, 38(9).
162. Medicare Australia. Australian childhood immunisation register (ACIR) statistics. 2013. www.medicareaustralia.gov.au.
163. White, C. England vaccination rates on the up but still below target. 2010. www.onmedica.com.
164. WHO. Immunisation position papers: Immunisation coverage. 2010. http://www.who.int/immunization/policy/en/index.html.
165. McIntyre, P., A. Williams, and J. Leask. Editorial—Refusal of parents to vaccinate: Deriliction of duty or legitimate personal choice? *Medical Journal of Australia*, 2003, 178: 160–161.
166. Sleet, D., B. Dinh-Zarr, and A. Dellinger. Traffic safety in the context of public health and medicine. In *Improving traffic safety culture in the United States: The journey forward*. 2007, AAA Foundation for Traffic Safety, Washington, DC.
167. Borkenstein, R.F., et al. *The role of the drinking driver in traffic accidents*. 1964, Department of Police Administration, Indiana University, Bloomington, IL.
168. McLean, A.J., O.T. Holubowycz, and B.L. Sandow. *Alcohol and crashes: Identification of relevant factors in this association*. 1980, Road Accident Research Unit, University of Adelaide, Adelaide, Australia.
169. CARRS-Q. *State of the road factsheet: Novice drivers*. 2011, Centre for Accident Research and Road Safety, Queensland, Australia.
170. Petroulias, T. *Community attitudes to road safety—2009 survey report*. 2009, Department of Infrastructure, Transport, Regional Development and Local Government, Canberra, Australia.
171. Victoria Police. *Victoria police annual report for 2011–2012*. 2012, Victoria Police, Melbourne, Australia.
172. Lane, J. Some recollections of the early days of road safety research. In *Speech following the Third Road Safety Researchers' Conference*. 1985, Melbourne, Australia.
173. Goldstein, N., S. Martin, and R. Cialdini. *YES! 50 secrets from the science of persuasion*. 2007, Profile Books, London.
174. Johnston, I.R. *Halving roadway fatalities: A case study from Victoria Australia 1989–2004*. 2006, U.S. Department of Transportation Federal Highway Administration, Washington, DC.
175. Lindblom, C., and E. Woodhouse. *The policy making process*, 3rd ed. 1993, Prentice Hall, Englewood Cliffs, NJ.
176. Crossen, C. *Tainted truth: The manipulation of fact in America*. 1996, Touchstone, New York.
177. Watson, D. Democracy for dummies. *The Age*, 2008.
178. Wiegmann, D., T. VonThaden, and A. Mitchell Gibbons. A review of safety culture theory and its potential application to traffic safety. In *Improving traffic safety culture in the United States: The journey forward*. 2007, AAA Foundation for Traffic Safety, Washington, DC.
179. Antonsen, S. Safety culture and the issue of power. *Safety Science*, 2009, 47: 183–191.

180. Perrow, C. *Normal accidents: Living with high-risk technologies*. 1984, Basic Books, New York.
181. Parker, D., M. Lawrie, and P. Hudson. A framework for understanding the development of organisational safety culture. *Safety Science*, 2006, 44: 551–562.
182. TAC. *Corporate statistics: Online crash database*. 2013. www.tac.vic.gov.au.
183. Australia, National Road Safety Council. *Monash university workshop: Essential occupational health and safety*. 2010, Monash University Accident Research Centre, Monash University, Melbourne, Australia.
184. Hauer, E. A case for evidence-based road-safety delivery. In *Improving traffic safety culture in the United States: The journey forward*. 2007, AAA Foundation for Traffic Safety, Washington, DC, pp. 329–344.
185. Hughes, B. *Workplace transport safety: The mobile financial crisis*. 2010, Curtin-Monash Accident Research Centre, Curtin University of Technology, Perth, Australia.
186. NTC. *A corporate approach to transport safety*. 2011, National Transport Commission, Melbourne, Australia.
187. Cameron, M., S. Newstead, and S. Gantzer. Effects of enforcement and supporting publicity programs in Victoria, Australia. In *International Conference, Strategic Highway Research Program (SHRP) and Traffic Safety*. 1995, Prague, Czech Republic.
188. Fahlquist, J. Responsibility ascriptions and Vision Zero. *Accident Analysis and Prevention*, 2006, 38: 1113–1118.
189. Elvik, R. Can injury prevention efforts go too far? Reflections on some possible implications of Vision Zero for road accident fatalities. *Accident Analysis and Prevention*, 1999, 31(3): 265–286.
190. Cloud, J. The myths of bullying. *Time Magazine*, 2012.
191. Robertson, L. *Injury epidemiology*. 1992, Oxford University Press, Oxford.
192. Kloeden, C.N., et al. *Travelling speed and the risk of crash involvement: Findings*. Volume 1. 1997, NHMRC Road Accident Research Unit, University of Adelaide, Australia.
193. Kloeden, C.N., A.J. McLean, and G. Glonek. *Reanalysis of travelling speed and the risk of crash involvement in Adelaide, South Australia*. 2002, Road Accident Research Unit, University of Adelaide, South Australia.
194. Kloeden, C., et al. *Travelling speed and the risk of crash involvement: Findings*. Volume 1. 1997, Federal Office of Road Safety, Canberra, Australia.
195. Hauer, E. *Speed and crash risk: An opinion*. 2004, RACV, Melbourne, Australia.
196. TRB. *A decade of experience—Special Report 204*. 1984, Transportation Research Board, Washington, DC.
197. Patterson, T., et al. The effect of increasing rural interstate speed limits in the United States. *Traffic Injury Prevention*, 2002, 3(4): 316–320.
198. Sligoris, J. 110 kilometre per hour speed limit—Evaluation of road safety effects. Quoted in Barton and Cunningham, Recent developments in speed management in Victoria. *The Speed Review: Appendix of Working Papers*. 1992, Monash University Accident Research Centre, Monash University, Melbourne, Australia.
199. Elvik, R. *The power model of the relationship between speed and road safety: Update and new analyses*. 2009, Institute of Transport Economics, Norwegian Centre for Transport Research, Oslo, Norway.
200. Newstead, S., N. Mullan, and M. Cameron. *Evaluation of the speed camera program in Victoria 1990—1993. Phase 5: Further investigation of localised effects on casualty crash frequency*. Monash University Accident Research Centre Report Series. 1995, Monash University Accident Research Centre, Melbourne, Australia.
201. Nilsson, G. *Traffic safety measures and observance: Compliance with speed limits, seat belt use and driver sobriety*. 2004, VTI Swedish National Road and Transport Research Institute, Linkoping, Sweden.

202. Nilsson, G. *Speeds, accident rates and personal injury consequences for different road types*. 1984, Swedish National Road and Transport Institute.
203. Taylor, M.C., A. Baruya, and J.V. Kennedy. *The relationship between speeds and accidents on rural single-carriageway roads*, ed. Department for Transport Road Safety Division, Local Government and the Regions. TRL Report TRL511. 2002, TRL.
204. Logan, D., and B. Corben. Speed risk curves: Unpublished road safety modelling. 2013, Monash University Accident Research Centre, Melbourne, Australia.
205. Wramborg, P. Fatality risk for three major crash types at different speed. In *Towards zero: Ambitious road safety targets and the Safe System approach*, ed. OECD. 2005, Transport Research Centre, Organisation for Economic Cooperation and Development, Paris.
206. OECD. Speed management. In *European Conference of Ministers of Transport*. 2006, Transport Research Centre, Organisation for Economic Cooperation and Development, Paris.
207. NHTSA. Estimating lives saved by electronic stability control 2009–2011. In *Traffic safety facts: Research notes*. 2013, National Highway Traffic Safety Administration, Washington, DC.
208. McLean, A., and R. Anderson. Metrication of the urban speed limit and pedestrian fatalities. In *Australian Road Safety Research, Policing and Education Conference*. 2008, Adelaide, Australia.
209. Job, S., et al. Community perceptions and beliefs regarding low level speeding and suggested solutions. In *TRB 92nd Annual Meeting Compendium of Papers*. 2013.
210. Austroads. *Driver attitudes to speed enforcement*. 2013, Austroads, Sydney, Australia.
211. Silcock, D., et al. What limits speed? Factors that affect how fast we drive. In *Final report: Summary and conclusions, June 2000*, ed. R.S.L.i.a.w.S.R. Associates. 2000, AA Foundation for Road Safety Research, UK.
212. Johnston, I. Reducing injury from speed related road crashes. *Injury Prevention*, 2004, 10: 257–259.
213. Reid, T. Voters put paid to speed cameras. *The Australian*, 2009.
214. Douma, F., L. Munnich, and T. Garry. *Identifying issues related to deployment of automated speed enforcement*. 2012, ITS Institute, University of Minnesota.
215. Gordon, J. Populist stance on speed cameras has left coalition in a jam. *The Age*, 2011.
216. VAGO. *Road safety camera program: Victorian Auditor-General's Report*. 2011, Victoria, Australia.
217. Donovan, R., et al. Self-regulation of motor vehicle advertising: Is it working in Australia? *Accident Analysis and Prevention*, 2011, 43: 631–636.
218. Schonfeld, C., D. Steinhardt, and M. Sheehan. A content analysis of Australian motor vehicle advertising: Effects of the 2002 voluntary code on restricting the use of unsafe driving themes. In *Australasian road safety research, policing and education*. 2005, Wellington, New Zealand.
219. Tate, A. Taking the racing line—Revvy or not? *Sydney Morning Herald*, 2011.
220. Moeckli, J., and J. Lee. *The making of driving cultures (research report)*. 2007, AAA Foundation for Traffic Safety, Washington, DC.
221. SWOV. *Sustainable safety*, ed. S.I.f.R.S. Research. 2006, Leidschendam, Netherlands.
222. Charlton, S., and P. Baas. *Speed change management for New Zealand roads*. 2006, Land Transport New Zealand.
223. AAP. Special issue on intelligent speed assist (ISA). *Accident Analysis and Prevention*, 2012, 48(special issue).
224. MacDonald, M., and S. Pope. Community safe speed promise: A joint initiative between the cities of Joondalup and Stirling. In *Australasian College of Road Safety National Conference*. 2009, Perth, Australia.

225. Kotter, J. Leading change: Why transformation efforts fail. *Harvard Business Review*, January 2007.
226. Lane, J. Death on the road: The prospects for control. *Medical Journal of Australia*, 1966, 2: 164–168.
227. Trinca, G. Media campaign for preventive action on the road toll. Personal communication with Ian Johnston. 1988.
228. McIntosh, L. *Road safety management in Australia: A call for more co-ordinated action*. 2013, Australasian College of Road Safety, Canberra.
229. Hinchcliff, R., et al. Media framing of graduated licensing policy debates. *Accident Analysis and Prevention*, 2010, 42: 1283–1287.
230. King, C. *Speech for Australian launch of Decade of Action for Road Safety 2011–20*, ed. M.f.I.a.T.M.f.R.D.a.L. Government. 2011.
231. Editorial. Politics the poorer when visions for nations are withheld: Abbott's lack of policy detail shows a lack of faith in voters. *Sydney Morning Herald National Times*, 2012.
232. Moeckli, J., and J. Lee. The making of driving cultures. In *Improving traffic safety culture in the United States: The journey forward*. 2007, AAA Foundation for Traffic Safety, Washington, DC.
233. Australasian College of Road Safety (ACRS). 2013. www.acrs.org.au.
234. IRAP. *Vaccines for roads*. 2012, iRAP, Hampshire, UK.
235. Larsson, T., N. Candappa, and B. Corben. *Flexible barrier systems along high speed roads: A life-saving opportunity*. 2003, Monash University Accident Research Centre, Melbourne, Australia.
236. Lowe, I. Learning from the elephants: Toward a rational future. In *A herd of white elephants? Some big technology projects in Australia*, ed. P. Scott. 1992, Hale & Iremonger, Sydney.
237. Osborne, D., and T. Gaebler. *Reinventing government: How the entrepreneurial spirit is transforming the public sector*. 1992, Addison-Wesley, Reading, MA.
238. FIA. *Road safety in France—Reflections on three decades of road safety policy*. 2006, FIA Foundation, London.
239. Wilson, T. No granny chic in nanny state shtick: Risk-averse paternalism makes for a perverse reversal of freedoms. *The Age*, 2012.
240. QUT. 2013. http://www.qut.edu.au/study/courses/graduate-certificate-in-road-safety.
241. Monash University Accident Research Centre. 2013. Melbourne, Australia.
242. Australian Transport Council. *National road safety strategy 2011–2020*. 2013.
243. Cameron, M. *Traffic enforcement strategy in Victoria Police's region one: Interim report*. 2008, Monash University Accident Research Centre, Melbourne, Australia.
244. Wegman, F., et al. *SUNflower+6: A comparative study of the development of road safety in the SUNflower+6 countries—Final report*. 2005, SWOV Institute for Road Safety Research, Leidschendam, Netherlands.
245. Organisation for Economic Cooperation and Development (OECD). Towards zero: Ambitious road safety targets and the Safe System approach. In *International Transport Forum*. 2008, Transport Research Centre, Paris.

Index

Milton Keynes UK
Ingram Content Group UK Ltd.
UKHW030900141024
449569UK00025B/1308